W9-CMN-032

TRANSLATIONS OF MATHEMATICAL MONOGRAPHS

VOLUME 74

The Linearization Method in Hydrodynamical Stability Theory

V. I. YUDOVICH

American Mathematical Society · Providence · Rhode Island

В. И. ЮДОВИЧ

МЕТОД ЛИНЕАРИЗАЦИИ В ГИДРОДИНАМИЧЕСКОЙ ТЕОРИИ УСТОЙЧИВОСТИ

ИЗДАТЕЛЬСТВО РОСТОВСКОГО
УНИВЕРСИТЕТА, 1984

Translated from the Russian by J. R. Schulenberger
Translation edited by Ben Silver

1980 *Mathematics Subject Classification* (1985 *Revision*). Primary 76D05, 75E05; Secondary 35Q10, 35K22.

ABSTRACT. The theory of the linearization method in the problem of stability of steady-state and periodic motions of continuous media is presented, and infinite-dimensional analogues of Lyapunov's theorems on stability, instability and conditional stability are proved for a large class of continuous media. Semigroup properties for the linearized Navier-Stokes equations in the case of an incompressible fluid are studied, and coercivity inequalities and completeness of a system of small oscillations are proved.

Library of Congress Cataloging-in-Publication Data

I͡Udovich, V. I. (Viktor Iosifovich)
[Metod linearizat͡sii v gidrodinamicheskoĭ teorii ustoĭchivosti. English]
The linearization method in hydrodynamical stability theory/V. I. Yudovich.
p. cm. – (Translations of mathematical monographs; v. 74)
Translation of: Metod linearizat͡sii v gidrodinamicheskoĭ teorii ustoĭchivosti.
Bibliography: p.
ISBN 0-8218-4528-4 (alk. paper)
1. Hydrodynamics. 2. Stability. 3. Navier-Stokes equations. I. Title. II. Series.
QA911.I9313 1989 89-315
532′.5–dc19 CIP

Contents

Introduction

In this book the basis of the linearization method (Lyapunov's first method) is expounded in application to the problem of steady-state and periodic motions of a viscous incompressible fluid and also to the problem of solutions of parabolic equations.

The first chapter is devoted to the Navier-Stokes equations linearized near some steady flow. It contains estimates of various norms of solutions of the homogeneous and inhomogeneous equations on an infinite time interval. These estimates, in particular, show that a solution of the homogeneous equations decays (or increases) exponentially with time, while the argument of the exponential function is related to the boundary of the real part of the spectrum of the corresponding steady-state operator. Some results of this chapter are subsequently applied in the proof of theorems of Lyapunov's first method for the the nonlinear Navier-Stokes equations. It is true that the first chapter contains much more material than is necessary for the subsequent exposition. In particular, the estimates of the leading derivatives of solutions of the Navier-Stokes system are almost never used further on—they can be applied to the investigation of stability in $W_p^{(l)}$; corresponding results can easily be obtained by the methods of Chapter II. We also remark that the proof of the solvability of boundary value problems on the basis of a priori estimates in this chapter is treated in a rather cursory manner, since this question has been well studied in [43].

Chapter I consists of seven sections.

In §1 a survey is given of some known methods of estimating integral operators in L_p spaces, and the interpolation theorems of M. Riesz and Marcinkiewicz are formulated, together with some simple corollaries of them. Next in this section a vector-valued generalization of the so-called extrapolation theorem of Simonenko is formulated, and it is shown how the results of Calderón and Zygmund on singular integral operators can be derived from it. The latter, as we know, are an appropriate tool for obtaining estimates in L_p of the leading derivatives of solutions of elliptic equations. Some such estimates for the case of the Poisson equation are presented at the end of §1, and it is shown how these estimates behave as $p \to 1, \infty$.

In §2 a number of estimates are established for solutions of the Cauchy problem for a linear ordinary differential equation in a Banach space with an operator A generating an analytic semigroup [37], [40]. Here the concept of an operator having fractional degree relative to A is introduced, and to estimate the values of such operators on solutions integral representations of the latter and interpolation theorems are used. The most precise estimates are obtained in the case where the operator A generates an analytic semigroup not in a single space but in an entire scale of L_p spaces. At the end of this section it is shown how the method of M. G. Kreĭn, developed in [38] for the case of a bounded operator, in combination with our method of deriving estimates, leads to theorems of the existence and uniqueness of periodic and almost periodic solution which are bounded in various norms.

The same class of evolution equations as in §2 are investigated in §3. Here it is a question of so-called coerciveness inequalities—estimates of the leading derivatives of solutions. We show that for such equations coercivity in L_p for some $p_0 > 1$ implies coercivity in L_p for all $p > 1$ (Theorem 3.1). An analogous assertion (without investigation of the behavior of the estimates for $p \to 1, \infty$) was formulated by Sobolevskiĭ in [72]. We further show that coerciveness inequalities for solutions of the Cauchy problem and for periodic solutions follow from one another (Theorem 3.2). Related to Theorem 3.2 is the assertion proved in the Appendix to §5 that estimates of the "leading derivatives" on a semi-infinite time interval in the L_p norm with a weight depending exponentially on time follow from coerciveness inequalities on a finite time interval. Theorem 3.2 can be applied to the Navier-Stokes equations in two ways. On the one hand, according to this theorem the L_p-estimates of the leading derivatives for the problem with initial data (Solonnikov [78], [79]) imply analogous estimates for the periodic problem. On the other hand, by applying Fourier series and Marcinkiewicz's theorem on multipliers to prove coerciveness inequalities in the case of the periodic problem, it is then possible with the help of Theorem 3.2 to derive coerciveness inequalities for the problem with initial data.

In the case where A is a selfadjoint operator in Hilbert space (or an operator obtained from a self-adjoint operator by a weak perturbation of the type of adding terms with lower order derivatives to an elliptic operator) coercivity for $p_0 = 2$ can be obtained directly, and then by Theorem 3.1 it holds also for any $p > 1$ (see Theorem 3.3; a somewhat more special assertion is formulated in [44]).

In §4 some applications of the results of §§1–3 to parabolic equations are given. M. Z. Solomyak [74] for the first boundary value problem and Agmon and Nirenberg [2] for a broad class of boundary conditions proved that an elliptic operator A_{2m} of order $2m$ generates an analytic semigroup in L_p. The fact that differential operators of order less than $2m$ have fractional degree relative to A_{2m} is derived from multiplicative inequalities. Therefore, the results of §2 are made concrete in the case of parabolic equations (in a bounded domain with sufficiently smooth boundary) in the form of sharp estimates of solutions of the problem with initial data in L_p and $L_{p,r} = L_r(0,\infty; L_p(\Omega))$.

Papers by Ladyzhenskaya, Solonnikov, and Sobolevskiĭ have been devoted to estimates in L_2 and L_p of solutions of parabolic equations (see the bibliography in [44]). The method applied in §4 is apparently new and well adapted to the needs of stability problems: the estimates are obtained directly on an infinite time interval with indication of the correct order of exponential growth (or decay) as $t \to +\infty$. We further remark that this method makes it possible to prove imbedding theorems for the function classes $W_p^{r,\dots,r,2mr}$ in the cylinder $\Omega \times [0,\infty)$. It would be desirable also to derive estimates of the leading derivatives of solutions of parabolic equations proceeding only from an estimate of the resolvent. So far no one has been able to do this, but, using the estimates of Slobodetskiĭ and Solonnikov in L_p, it is possible to estimate derivatives in $L_{p,r}$ with the help of Theorem 3.1.

The study of the Navier-Stokes equations linearized in a neighborhood of some steady-state solution, which is henceforth called the *basic solution*, is begun in §5 of Chapter 1. The flow region Ω is assumed to be bounded. The subspaces S_p and G_p in the space of vector fields of class L_p on Ω obtained by closing the set of smooth solenoidal vector fields with zero normal component, and hence the set of gradients of smooth functions, are introduced at the beginning of this section.

It is proved that the operator Π of orthogonal projection in L_2 onto S_2 admits continuation from M to a continuous projection onto S_p in L_p for $p > 1$, and the space L_p is decomposed into a direct sum of the subspaces S_p and G_p. Application of the operator Π to the Navier-Stokes equations makes it possible to eliminate the pressure and drop the condition that the velocity field be solenoidal, which is a consequence of the equation obtained in this manner. The operator Π was used to this end by E. Hopf [24] and S. G. Kreĭn [39]. We note that this manner of eliminating the pressure is equivalent to that applied beginning with Hopf and Ladyzhenskaya (see [43]) in defining generalized solutions by means of integral identities.

In §5 it is further proved that the set of smooth solenoidal vectors vanishing on the boundary is dense in S_p (in the case $p = 2$ this assertion was proved in another manner in [43]), the general form of a continuous linear functional in S_p is indicated, and closedness and discreteness of the spectrum of the steady-state operator linearized about a state of rest of the Navier-Stokes operator in S_p are proved (regarding these results in the case $p = 2$, see [43]). It is shown (Lemma 5.6) that the spectrum of small oscillations about a given steady-state regime is discrete and that the sequence of eigenvectors and associated vectors is complete in M_0 in the metric of $W_p^{(2)}$ $(p > 1)$ (and hence in L_p and $W_p^{(1)}$). The proof is based on a result of Keldysh ([30]; see also [20]). An assertion regarding completeness in L_2 is contained in the work of S. G. Kreĭn's paper [39], where the possibility of applying Keldysh's theorem to this problem was apparently first indicated. Lemma 5.6 also contains an estimate of the eigenvalues from which it follows that for the given basic flow there can exist no more than a finite number of modes increasing with time.

An estimate in L_p of the resolvent of the linearized steady-state Navier-Stokes operator is derived in §5, and it is shown (Theorem 5.1) that in S_p it generates an analytic semigroup. By applying the general theorems of the preceding sections, we see that this makes it possible to obtain estimates of the lower derivatives of solutions of the problem with initial data, and also to derive estimates in $L_{p,r}$ from estimates in L_p of the leading derivatives (for the problem with initial data as well as for the periodic problem).

The semigroup approach to the investigation of solvability of the linear and nonlinear Navier-Stokes equations was applied by Sobolevskiĭ in [73]. This was preceded by the investigations of Krasnosel′skiĭ, S. G. Kreĭn, Sobolevskiĭ, and Kato, where abstract differential equations in Hilbert and Banach spaces were considered and fractional powers for various classes of linear operators were studied. (For a bibliography, [37], [43], and [40].)

The results of §§6 and 7 are used in the considerations of §5; we now proceed to an exposition of them.

Estimates of the resolvent of the steady-state Navier-Stokes operator linearized about a state of rest as derived in §6. Theorem 6.1 proved here asserts that M. Z. Solomyak's condition [74] is satisfied for this operator, and it generates an analytic semigroup on S_p for $p > 1$. This assertion is also contained in Sobolevskiĭ's paper [73]; he showed that it follows from estimates for the time-dependent problem with initial data [78] or the periodic problem [84] (see also §7 of Chapter I). An immediate consequence of this theorem is given in §6. To prove it in the case of a half-space we use a representation of the solution in the form of a Fourier integral, and an estimate is obtained by applying Mikhlin's theorem on multipliers of Fourier integrals and a theorem on the boundedness of the Hilbert integral transform in L_p. With the help of the usual technique of rectifying the boundary it is possible after this to go over to an arbitrary domain. Another method of estimating the solution of the steady-state boundary value problem, not requiring preliminary investigation of the problem in a half space, was given by Solonnikov and is expounded in [43]. This method, which is based on application of hydrodynamic potentials and reduction of the problem to an integral equation on the boundary of the domain, is especially convenient for the derivation of sharp estimates in the case where the right side is a generalized vector-valued function of the type of a derivative of a vector-valued function in L_p (Theorem 6.2). At the end of §6 we show that the Navier-Stokes operator generates an analytic semigroup in $S_p^{(1)}$ (the closure of M_0 in $W_p^{(1)}$).

At the beginning of §6 lemmas are presented which contain estimates in L_p of solutions of the familiar problem on the recovery of a solenoidal vector with normal component equal to zero on the boundary on the basis of a given vorticity and circulations over all independent contours. In the case $p = 2$ these estimates are known (K. Friedrichs, È. B. Bykhovskiĭ, N. V. Smirnov, and others; see [43]).

Coerciveness inequalities in L_p for the Navier-Stokes equations for the problem with initial data were proved by Solonnikov [77], [78] using time-dependent hydrodynamic potentials. It would be very desirable to learn to derive them from estimates of the resolvent. As mentioned above, this has so far not been possible. They can nevertheless be obtained by repeating for time-dependent equations the same arguments as in the steady-state case. This is done in §7 first for flows periodic in time (Theorem 7.1) and then for the problem with initial data with application of Theorem 3.2 (Theorem 7.3). We further note that Solonnikov gave a detailed investigation of conditions on the initial and boundary data for the Navier-Stokes system which guarantee the possibility of an extension with prescribed differential properties, and he also proved coerciveness inequalities in Hölder spaces. We do not touch on these questions here; we mention only that, on the basis of the results of Solonnikov [77], [78], by the methods indicated below it is not hard to investigate stability in Hölder spaces.

In Chapter II we present an infinite-dimensional generalization of Lyapunov's first method (= the method of linearization = the method of small oscillations) to a certain class of evolution equations which include the Navier-Stokes equations, parabolic equations, the equations of magnetohydrodynamics, etc.; the stability of steady and time-periodic motions is considered.

The first work on the generalization of Lyapunov's stability theory to the infinite-dimensional case was done back in the forties by M. G. Kreĭn, who considered ordinary differential equations in a Banach space with continuous operators on the right side. The basic difficulty occasioned by infinite-dimensionality and overcome by Kreĭn here consists in the more complex structure of the spectrum of the linear operator, which makes it necessary to give up many methods of linear algebra on which the investigation in the theory of ordinary differential equations is based. The results of these investigations are expounded in the monographs [38] and [12].

Of course, a solution of some questions of stability theory for equations with unbounded operators, in particular, for equations of mathematical physics, an be obtained by the same methods as for equations with continuous operators. However, stability theory for nonlinear partial differential equations has a number of special features.

For parabolic equations and the Navier-Stokes equations the Cauchy problem has a well-defined solution only for positive times; the operator of translation along trajectories for the corresponding linear equations turns out not to be invertible. The presence of strong nonlinearities in the equations (say, a dependence of the nonlinear terms on the derivatives of the unknown functions) makes it necessary to estimate various norms of solutions of the homogeneous and inhomogeneous equations (see Chapter I); the model of an equation in a Banach space turns out to be too restrictive—in many cases it is unnatural to assume that the derivative of the unknown vector-valued function belongs to the same space as the function itself (see Theorems 2.1 and 2.2). Quite generally, the choice of space in which stability is investigated turns out to be a rather critical matter for specific equations: in §1 of Chapter II we present examples of partial differential equations for which stability holds in certain spaces and fails in others for the same motion. This situation occurs, for example, for the Euler equations linearized about two-dimensional Couette flow in a channel: the vorticity of the perturbation remains bounded, while its derivatives grow unboundedly with time.

The vorticity of a perturbation of Couette flow is thus reminiscent over a large time of a continuous and nowhere differentiable Weierstrass function.

We remark that at the present time such behavior has been discovered for a rather larger class of solutions of nonlinear equations of the motion of an ideal incompressible fluid. In the three-dimensional case the velocity vector remains bounded, while the vorticity increases (see [85]).

It is entirely possible that the properties of the loss of smoothness of a solution which we have described play an essential role in the development of turbulence: a system described by a continuous but nowhere differentiable function must behave like a stochastic system.

In §1 of Chapter II various definitions of the stability of motion of an infinite-dimensional system are considered. For what follows (exponential asymptotic) Lyapunov stability in a Banach space and also stability in mean play an essential role. In some cases it is convenient to estimate the perturbation at the initial time $t = 0$ in the norm of a space X, while for $t > 0$ it is convenient to estimate it in the norm of Y (stability from the space X to the space Y). The concept of η-stability is convenient to describe progressive smoothness (with increasing time) of solutions of parabolic equations and the Navier-Stokes equations.

Analogues of Lyapunov's theorem regarding asymptotic stability for a nonlinear ordinary differential equation in Banach space (Theorems 2.1 and 2.2) are proved in §2. Specialization of these theorems for the case of the Navier-Stokes equations leads to Theorems 2.3 and 2.4. The latter assert that in the case where the spectrum of the linearized problem (about a given steady flow) is situated in the left half-plane the basic flow is exponentially asymptotically stable in the Lyapunov sense in the metric of L_p for any $p \geq 3$ and is also exponentially stable in the mean in various metrics. It is not without interest to note that the divergent character of the nonlinear terms in the Navier-Stokes equations is of basic significance in the derivation of the sharp estimate in the case $p = 3$ contained in Theorem 2.4.

Theorems 2.1 and 2.2 can be applied to various parabolic problems. This possibility is illustrated by Theorem 2.5, which provides the justification of the legitimacy of linearization for a parabolic equation with a nonlinear term depending in power fashion on the derivatives of the unknown function.

In §3 it is proved for the same class of equations as in §2 that existence for the linearized problem of points of the spectrum in the right half-plane entails instability. Moreover, theorems on the existence (in the case where the spectrum of the linearized problem does not contain points of the imaginary axis) of invariant manifolds which are stable as $t \to +\infty$ and $t \to -\infty$ (conditional stability) are proved.

For ordinary differential equations stable invariant manifolds under various assumptions were investigated by A. M. Lyapunov, I. G. Petrovskiĭ, O. Perron, N. N. Bogolyubov, Yu. A. Mitropol′skiĭ, O. B. Lykova, and others (the last two authors also studied the question in the case of Hilbert space, while Daletskiĭ did the same for a Banach space). A bibliography of these works is contained in [58] and [12]. The case of a differential equation with a continuous nonlinearity in a Banach space is considered in the book by Daletskiĭ and Kreĭn [12]. A corresponding result for the Navier-Stokes equations was reported by the author at the Second All-Union Conference on Theoretical and Applied Mechanics (Moscow, 1964) and was published in [90].

In the works listed above the investigation is based, as a rule, on the reduction of the problem with initial data to an integral equation on a semi-infinite time interval by inverting the leading linear part of the differential equation in question. Another approach to the stability problem is related to the study of iterations of the operator of translation along trajectories of the differential equation in question. This approach stems from Hadamard and is applied in a number of works, of which we mention Anosov's paper [4] (see also [5]), in which finite-dimensional case is considered, and also Neĭmark's papers [56] and [57], where differential equations with continuous right sides in a Banach space are considered. In Krasnosel′skiĭ's book [36] the translation operator is used in the investigation of periodic motions by conic and topological methods.

In Chapter III the translation operator is applied to investigate stability of periodic motions of a fluid (a new method of proof for steady flows is obtained at the same time). With the desire of making the exposition in this chapter independent of the preceding chapter, we have restricted ourselves to an investigation of stability in an energy space. The required properties of the translation operator are mainly known and are presented in the required form in Lemmas 2.1–2.5. We remark that the operator Lemmas 3.1, 4.2, and 5.1 in Chapter 3 in combination with known estimates of solutions of the Navier-Stokes system (see [43] and also Chapter I) make it possible in an entirely analogous manner to study stability in other function spaces (for example $W_p^{(l)}$ and $C^{K,\lambda}$) in which at least local solvability of the problem with initial data holds.

A general existence theorem for periodic flow is contained in [85], and in the two-dimensional case also in [64] and [66].

Theorems 3.1 and 4.1 of Chapter III make it possible to ascertain stability and, correspondingly, instability of a periodic regime on the basis of the spectrum of the linearized problem (their proofs are based on the operator Lemmas 3.1 and 4.1).

In the special case of steady flows Theorem 3.1 was proved by Prodi [62], who applied the method of integral equations. Stability in the class of weak generalized solutions was investigated by Sattinger [64].

Theorem 5.1 contains a detailed description of the behavior of the trajectories in a neighborhood of an unstable periodic regime in the hyperbolic case when the spectrum of stability does not contain points of the imaginary axis while the spectrum of the monodromy operator does not contain points of the unit circle. The corresponding operator Lemma 5.1 is proved basically by the method of [4], which is also used in [57]; a new circumstance is the noninvertibility of the translation operator along trajectories. The latter is completely continuous.

Lemma 5.2 is next proved; it asserts that the nonzero spectrum (and also the nonzero point, continuous, and residual spectrum) of a product of two linear operators acting in some Banach space does not depend on the order of the factors. With the help of this lemma it is proved that stability of a periodic regime does not depend on the choice of the initial time.

Theorems 8.1 and 8.2 of §8 assert that for smooth data in the case of asymptotic stability perturbations die out in time together with all their derivatives.

Lemma 6.1 in §6 is an infinite-dimensional analogue of the Andronov-Vitt theorem on the stability of a self-oscillatory periodic regime—a cycle. The corresponding result for the Navier-Stokes equations (Theorem 6.1) also follows from it. Theorem 7.1 contains a condition for instability of a cycle as an invariant manifold; a simpler assertion regarding the instability of individual periodic regimes follows from Theorem 4.1. Finally, Lemma 7.2 describes the behavior of trajectories in a neighborhood of an unstable hyperbolic cycle and establishes the existence of stable and unstable invariant manifolds of a cycle.

The results presented in Chapter III were published in the author's papers [100] and [101] (see also [86]). We remark that some critical cases of instability for equations of parabolic type were considered by Kolesov [32]. An infinite-dimensional generalization of the Andronov-Vitt theorem different from that given in Chapter III is contained in [12] (Chapter VII, §4.4, Theorem 4.5; we note that in this theorem the superfluous condition that its spectrum not encircle zero is imposed on the monodromy operator).

At the present time the linearization method is basic in the theory of stability of fluid flow—in any case it has been possible to detect instability only by this method. Many papers have been devoted to an analysis of stability of specific flows on the basis of an investigation of the spectrum of the linearized system (see the books by Lin [45], Monin and Yaglom [55], Joseph [29], and also Gol′dshtik and Shtern [105]).

The Navier-Stokes equations linearized about a steady-state flow turn out to be non-self-adjoint [99]. Their investigation is a difficult problem, and for the solution of the majority of questions it is necessary to have recourse to numerical methods. An analytic consideration is possible only in a few cases. The arsenal of methods applicable here is not large. Sometimes the linearized operator nevertheless turns out to be self-adjoint or symmetrizable [99] but, it is true, only for perturbations with special symmetry. In some cases it is possible to apply the theory of positive operators [34], [35]. The theory of oscillation operators [18] is used in an essential manner in the study of change of stability in the problem of the occurrence of Taylor vortices [95], [96].

In cases where the basic regime contains only one spatial or time harmonic, reduction to a tridiagonal Jacobi system in the space of Fourier coefficients provides assistance. In such cases the secular equation admits a representation in terms of continuous fractions [48], [49], [51], [91]). Various asymptotic and perturbation methods are also widely used (see, for example, [41], [42], [94], and [101]–[103]).

A new method of spectral theory has been developed by Yu. S. Barkovskiĭ and the author [106].

Among the classical problems which still await solution we mention the justification of the "principle of change of stability" for rotational flows [45] and a rigorous proof of the stability of Poiseuille flow in a circular pipe and of Couette flow in a channel for all Reynolds numbers. For the latter flow a result with a minimal use of numerical methods was obtained by Romanov [63]. These problems merit attention in particular because their solution in all probability requires new methods in the spectral theory of differential operators.

The author is grateful to I. I. Vorovich, under whose supervision he began to concern himself with mathematical hydrodynamics and the theory of stability, and also to V. I. Arnol′d, D. V. Anosov, M. A. Krasnosel′skiĭ, S. G. Kreĭn, Yu. P. Krasovskiĭ, O. A. Ladyzhenskaya, I. B. Simonenko, and V. A. Solonnikov for useful and stimulating discussions.

CHAPTER I

Estimates of Solutions of the Linearized Navier-Stokes Equations

§1. Estimates of integral operators in L_p

In this section we give a survey of some modern methods of estimating linear operators, in particular, integral operators, in L_p. Ideas of interpolation and extrapolation of operators play an important role below.

We denote by (Ω, μ) a measure space, i.e., a set Ω with a distinguished σ-algebra of subsets Σ on which there is defined a countably additive non-negative set function μ (it is not exluded that $\mu(A) = +\infty$ for some sets $A \in \Sigma$).

$L_p(\Omega, \mu, X)$ for $1 \leq p < \infty$ denotes the Lebesgue space of μ-measurable functions(1) defined on Ω with values in a Banach space X. The norm in $L_p(\Omega, \mu, X)$ is defined by

$$\|f\|_p = \|f\|_{L_p(\Omega,\mu,x)} = \left[\int_\Omega \|f(\omega)\|_X^p \, d\mu(\omega)\right]^{1/p}. \tag{1.1}$$

It is known [15] that $L_p(\Omega, \mu, X)$ is a Banach space. It is frequently convenient to consider any Banach space as the closure of its "best representers." It is essential that each of the spaces $L_p(\Omega, \mu, X)$ is the closure of the same set $L^0(\Omega, \mu, X)$ of all μ-simple functions. A function $f(x)$ is called *μ-simple* if it assumes a finite number of values $f_1, \dots, f_k$ and the sets $F_i = \{x \colon x \in \Omega; f(x) = f_i\}$ $(i = 1, \dots, k)$ are μ-measurable, with $\mu(F_i) < \infty$ if $f_i \neq 0$.

In the case where Ω is n-dimensional Euclidean space R^n, μ is Lebesgue measure, and $X = R^1$ the set of compactly supported smooth functions(2) possesses the same property.

(1)Strictly speaking, classes of μ-equivalent functions ($f \sim g$ if $\mu(\{x \colon f(x) \neq g(x)\}) = 0$). It is well known that this ambiguity leads to no misunderstandings.

(2)The word "smooth" is here and henceforth used in the sense of "sufficiently smooth or, if appropriate, infinitely differentiable."

Below we shall consider linear operators taking one Lebesgue space into another, and we shall be interested in conditions for their boundedness. Keeping in mind the fact that a bounded operator defined on a dense set of a Banach space can be extended to the whole space by continuity, it suffices to define operators on L^0. In particular, this makes it possible to consider a linear operator $A: L_{p_0}(\Omega, \mu, X) \to Y$ (Y is a Banach space) as a linear operator from $L_p(\Omega, \mu, X)$ to Y ($1 \le p < \infty$) which is possibly unbounded and defined not everywhere but only on a dense set.

1. Interpolation theorems. The forefather of all presently known interpolation theorems is the following theorem of M. Riesz (see [15], [37] and [104]).

THEOREM 1.1 (the M. Riesz interpolation theorem). *Suppose a linear operator A acts boundedly from $L_{p_1}(\Omega_1, \mu_1, R^1)$ to $L_{q_1}(\Omega_2, \mu_2, R^1)$ and from $L_{p_2}(\Omega_1, \mu_1, R^1)$ to $L_{q_2}(\Omega_2, \mu_2, R^1)$:*

$$A: L_{p_1}(\Omega_1, \mu_1, R^1) \to L_{q_1}(\Omega_2, \mu_2, R^1); \qquad (1.2)$$
$$A: L_{p_2}(\Omega_1, \mu_1, R^1) \to L_{q_2}(\Omega_2, \mu_2, R^1).$$

Then

$$A: L_p(\Omega_1, \mu_1, R^1) \to L_q(\Omega_2, \mu_2, R^1); \qquad (1.3)$$
$$\frac{1}{p} = \frac{\alpha}{p_1} + \frac{1-\alpha}{p_2}; \; \frac{1}{q} = \frac{\alpha}{q_1} + \frac{1-\alpha}{q_2},$$

and

$$\|A\|_{p,q} \le c\|A\|^{\alpha}_{p_1,q_1}\|A\|^{1-\alpha}_{p_2,q_2}; c = c(p_1, p_2, q_1, q_2, \alpha). \qquad (1.4)$$

Here we have introduced the notation

$$\|A\|_{p,q} = \sup_{f \in L^0(\Omega_1, \mu_1, R^1)} \frac{\|Af\|_{L_q(\Omega_2, \mu_2, R^1)}}{\|f\|_{L_p(\Omega_1, \mu_1, R^1)}}. \qquad (1.5)$$

In the case of complex L_p the constant c in (1.4) can be omitted.

THEOREM 1.2 (the Marcinkiewicz interpolation theorem [16], [37]). *Suppose that A is a linear operator mapping $L^0(\Omega_1, \mu_1, X_1)$ into a set of μ_2-measurable functions defined on Ω_2 with values in X_2. Suppose that for all $a > 0$*

$$[\mu_2(\{\omega_2 : |(Af)(\omega_2)|x_2 > a\})]^{1/q_k} \le \frac{M_k}{a}\|f\|_{L_{p_k}(\Omega_1, \mu_1, X_1)} \qquad (1.6)$$

with constants μ_1 and μ_2 not depending on f. Then

$$A: L_p(\Omega_1, \mu_1, X_1) \to L_q(\Omega_2, \mu_2, X_2);$$
$$\left(\frac{1}{p} = \frac{\alpha}{p_1} + \frac{1-\alpha}{p_2}; \quad \frac{1}{q} = \frac{\alpha}{q_1} + \frac{1-\alpha}{q_2}; \quad 0 < \alpha < 1\right) \qquad (1.7)$$

and

$$\|A\|_{p,q} \leq \frac{2(\frac{1}{q_1} - \frac{1}{q_2})}{(\frac{1}{q} - \frac{1}{q_2})(\frac{1}{q_1} - \frac{1}{q})} M_1^{\alpha} M_2^{1-\alpha}. \tag{1.8}$$

Marcinkiewicz's theorem considerably strengthens Riesz's theorem (true, it is necessary to impose the additional condition $q_k \geq p_k$). Indeed, it is obvious that conditions (1.2) imply (1.6) but not conversely. Further improvements of the interpolation theorems are known. Thus, Stein and Weiss [37], [80] showed that it suffices to verify condition (1.6) (for $X_1 = X_2 = R^1$) for characteristic functions of measurable sets.

Dikarev and Matsaev in [14] established that in conditions (1.2) (and hence also in (1.6)) the operator acts into a function space considerably smaller than L_q.

We shall present some applications of the Marcinkiewicz theorem. We remark that direct application of this theorem is especially simple when the operator A is such that the function Af is good everywhere except for some known singular points. For example, it is often convenient to apply the following simple assertion.

THEOREM 1.3. *In order that a linear operator act boundedly from* $L_p(\Omega, \mu, X)$ *to* $L_q(R^n, m_n, Y)$ ($1/q = \beta/p - \gamma; 0 \leq \gamma \leq \beta - 1; (\beta - 1)/\gamma < p < \beta/\gamma$; *$\beta$ and γ do not depend on p, and m_n is n-dimensional Lebesgue measure*) *it suffices that the inequality*

$$|(Af)|_Y \leq c|x|^{-n/q} \|f\|_{L_p(\Omega,\mu,X)} \tag{1.9}$$

be satisfied for all $f \in L^0(\Omega, \mu, X)$ and $x \in R^n$ with the same constant c.

PROOF. From (1.9) it follows that

$$\begin{aligned} [m_n(\{x\colon |(Af)(x)|_Y > a\})]^{1/q} &\leq [m_n(\{x\colon c|x|^{-n/q}\|f\|_{L_p} > a\})]^{1/q} \\ &= \omega_n^{1/q} \cdot \frac{c}{a} \|f\|_{L_p}, \end{aligned} \tag{1.10}$$

where ω_n is the volume of the n-dimensional unit ball. It remains to apply the Marcinkiewicz theorem, and Theorem 1.3 is proved.

THEOREM 1.4. *The Hilbert integral operator*

$$(Af)(s) = \int_0^\infty \frac{f(t)}{t+s}\, dt \tag{1.11}$$

acts boundedly in $L_p((0, \infty), m_1, X)$.

PROOF. Applying the Hölder inequality, from (1.11) we deduce that

$$\begin{gathered} |(Af)(s)|x \leq \int_0^\infty \frac{|f(t)|}{t+s}\, dt \leq \|f\|_{L_p} \cdot C_p s^{-1/p}; \\ C_p = (p-1)^{-1/p}. \end{gathered} \tag{1.12}$$

The boundedness of the operator A in L_p for any $p > 1$ now follows from Theorem 1.3, as required.

2. An extrapolation theorem and singular integrals. Calderón and Zygmund in [10] proved boundedness in L_p of singular integral operators with difference kernels. In analyzing their proof, I. B. Simonenko noted that their method leads to the result whenever the operator acts in L_p (for some $p_0 > 1$) and a certain (easily verified) estimate of the kernel is satisfied. In particular, it is unimportant that the kernel be a difference kernel.

The corresponding theorem, called an extrapolation theorem, was formulated in [67]–[69], where a generalization of it to Orlicz spaces was also given. Below we present the extrapolation theorem for the vector case and derive the result of Calderón and Zygmund from it. The method of proof follows J. Schwartz [69], where a more special result is obtained.

THEOREM 1.5. *Suppose $K(x,y)$ is a function defined on $R^n \times R^n$ with values in the space of bounded linear operators ($X \to Y$), where X and Y are Banach spaces. On the set of functions $f \in L^0(R^n, m_n, X)$, define an operator A by setting*

$$(Af)(x) = \int_D K(x,y)f(y)\,dy, \tag{1.13}$$

where D is any measurable set. Suppose A acts boundedly from $L_p(D, m_n, X)$ to $L_r(R^n, m_n, Y)$, and

$$\int_{S'_{2h}} |K(x,y) - K(x,y')|^q\,dx \le B^q, \tag{1.14}$$

where $q \ge 1$ and $1/p - 1/r = 1 - 1/q$; S_h and S_{2h} are concentric cubes with sides h and $2h$ respectively; $y, y' \in S_h$, and the integration goes over the complement to the larger cube S_{2h}. Here the constant B need not depend on h, y, y', or the choice of the center of the cube. Then the operator A acts boundedly from L_{p_1} to L_{r_1} if $1 < p_1 \le p$; $1/p_1 - 1/r_1 = 1 - 1/q$. Moreover,

$$\|A\|_{p_1,r_1} \le \frac{B_1}{r_1 - 1} \tag{1.15}$$

with a constant B_1 depending only on $\|A\|_{p,r}$ and B.

PROOF. For any $t > 0$ it suffices to obtain the estimate

$$[m_n(\{x \colon x \in R^n; |(Af)(x)| > t\})]^{1/q} \le \frac{C}{t}\|f\|_{L_1} \tag{1.16}$$

and apply the Marcinkiewicz interpolation theorem. We use the following lemma of Calderón and Zygmund in a form given by Hörmander ([25], Lemma 2.2).

LEMMA 1.1. *Suppose $s > 0$ and $u \in L_1(R^n, m_n, X)$. Then the function $u(x)$ can be represented in the form*

$$u = v + \sum_{k=1}^{\infty} w_k, \tag{1.17}$$

and the following conditions are satisfied:

$$|v|_1 + \sum_{k=1}^{\infty} |w_k|_1 \leq 3|u|_1, \tag{1.18}$$

$$|v(x)| \leq 2^n s \tag{1.19}$$

for almost all $x \in R^n$; the function w_k vanishes outside a cube I_k with side length 2^{-n_k} (n_k is a natural number); these cubes do not intersect, $I_k I_l = \varnothing$ $(k \neq l)$, and

$$\sum_{k=1}^{\infty} m_n(I_k) \leq \frac{1}{s}\|u\|_1; \tag{1.20}$$

$$\int_{I_k} w_k(x)\, dx = 0. \tag{1.21}$$

A proof of this lemma can be found in [10] or [25].

Below we use the notation

$$(f)_t = m_n(\{x \colon |f(x)| > t\}). \tag{1.22}$$

It is clear that

$$(Au)_t \leq (Av)_{t/2} + (Aw)_{t/2}; w = \sum_{k=1}^{\infty} w_k. \tag{1.23}$$

We now estimate $(Av)_t$ and $(Aw)_t$. According to the Tchebycheff inequality,

$$(Av)_t \leq t^{-r}\|Av\|_r^r. \tag{1.24}$$

Using the boundedness of $\|A\|_{p,r}$ and the estimates (1.18) and (1.19), from (1.24) we further obtain

$$(Av)_t \leq t^{-r}\|A\|_{p,r}^r \cdot \|v\|_p^r \leq C_1 t^{-r} s^{r(1-1/p)} \|u\|_1^{r/p}; \tag{1.25}$$
$$C_1 = 3^{r/p}\|A\|_{p,r} \cdot 2^{nr(1-1/p)}.$$

By (1.21) we have

$$(Aw_k)(x) = \int_{I_k} [K(x,y) - K(x,y')] w_k(y)\, dy. \tag{1.26}$$

Suppose I_{k_1} is a cube concentric with I_k and with double side length. Using (1.24), from (1.26) we deduce that

$$\|Aw_k\|_{L_q(I'_{k_1})} \leq \int_{I_k} |w_k(y)|x \cdot \|K(x,y) - K(x,y)\|_{L_q(I'_{k_1})}\, dy \\ \leq B\|w_k\|_1. \tag{1.27}$$

Let $e = \bigcup_1^\infty I_{k1}$. Then by (1.20)

$$m_n(e) \leq \sum_{k=1}^{\infty} m_n(I_{k1}) \leq 2^n \sum_{k=1}^{\infty} m_n(I_k) \leq \frac{2^n}{s}\|u\|_1. \tag{1.28}$$

We now note that

$$(Aw)_t = m_n(\{x\colon |(Aw)(x)| > t\} \cap e) + m_n(\{x\colon |(Aw)(x)| > t\} \cap e') \\ \leq m_n(e) + t^{-q}\|Aw\|^q_{L_q(e')}. \tag{1.29}$$

We estimate the second term in (1.9) by means of (1.27) and (1.18):

$$\|Aw\|_{L_q(e')} \leq \sum_{k=1}^{\infty} \|Aw_k\|_{L_q(e')} \leq B \sum_{k=1}^{\infty} \|w_k\|_1 \leq 3B\|u\|_1. \tag{1.30}$$

With the help of (1.28) and (1.30), from (1.29) we now obtain

$$(Aw)_t \leq \frac{2n}{s}\|u\|_1 + (3B)^q t^{-q}\|u\|_1^q. \tag{1.31}$$

Further, estimating the right side in (1.23) with the help of (1.25) and (1.31), we find that

$$(Au)_t \leq C_1 2^{-r} t^{-r} s^{r(1-1/p)} \|u\|_1^{r/p} + \frac{2^n}{s}\|u\|_1 + (3B)^q t^{-q}\|u\|_1^q. \tag{1.32}$$

It is now time to use the arbitrariness in the choice of the parameter s. In order that the right side in (1.32) be minimal it is necessary to set

$$s = C_2 t^q \|u\|_1^{1-q}; C_2 = \left[2^{n+r}/C_1\left(\frac{r}{q} - 1\right)\right]^{q/r}. \tag{1.33}$$

The condition $1/p - 1/r = 1 - 1/q$ has been used here. From (1.32) we now obtain

$$(Au)_t^{1/q} \leq \frac{C}{t}\|u\|_1; \\ C = \left[2^{-r} C_1 C_2^{r(1-1/p)} + \frac{2^n}{C_2} + (3B)^q\right]^{1/q}, \tag{1.34}$$

which coincides with (1.16). The estimate (1.15) follows from (1.8). The theorem is proved.

COROLLARY 1. *Suppose the transposed kernel $K(y,x)$ satisfies a condition of the form* (1.14), *i.e.,*

$$\int_{S'_{2h}} |K(x',y) - K(x,y')|^q \, dx \le B_1^q. \tag{1.35}$$

The operator A in the conditions of Theorem 1.5 *then acts boundedly from L_p to L_r if*

$$p_1 > p; \frac{1}{p_1} - \frac{1}{r_1} = 1 - \frac{1}{q}. \tag{1.36}$$

PROOF. It is known [15] that the adjoint operator A^* is

$$(A^* f)(x) = \int_{R^n} K^*(y,x) f(y) \, dy. \tag{1.37}$$

Applying the theorem just proved to the operator (1.37) and noting that

$$\|A\|_{L_p(\Omega,\mu,X)\to L_p(\Omega,\mu,X)} = \|A\|_{L_{p'}(\Omega,\mu,X^*)\to L_{p'}(\Omega,\mu,X^*)},$$

we obtain Corollary 1.

COROLLARY 2. *Condition* (1.14) *of Theorem* 1.5 *can be replaced by*

$$h^q \int_{S'_{2h}} |\nabla_y K(x,y)|^q \, dx \le B^q, \tag{1.38}$$

where $h > 0$ is arbitrary, $y \in S_h$, and B does not depend on h or y.

PROOF. We shall show that (1.14) follows from (1.38). For $y, y' \in S_h$ and $z = y' - y$ we have

$$\begin{aligned} |K(x,y) - K(x,y')|^q &= \left| \int_0^1 \sum_{k=1}^n \frac{\partial K(x,y+tz)}{\partial y_k} z_k \, dt \right|^q \\ &\le h^q \int_0^1 |\nabla_y K(x,y+tz)|^q \, dt. \end{aligned} \tag{1.39}$$

Noting that $y + tz \in S_h$, from (1.38) and (1.39) we conclude that

$$\int_{S'_{2h}} |K(x,y) - K(x,y')|^q \, dx \le h^q \int_0^1 dt \int_{S'_{2h}} |\nabla_y K(x,y+tz)|^q \, dx \le B^q. \tag{1.40}$$

Corollary 2 is thus proved.

THEOREM 1.6 (Calderón and Zygmund [10]). *The singular integral operator*

$$(Af)(x) = \int_{R^n} \frac{\omega(x-y)}{|x-y|^n} f(y) \, dy, \tag{1.41}$$

where $\omega(z)$ is a homogeneous function of degree zero, is smooth, and has an integral over the unit sphere equal to 0, acts boundedly in $L_p(R^n)$ for any $p > 1$. Moreover,

$$\|A\|_{L_p \to L_p} \leq \frac{Cp^2}{p-1},$$

where C does not depend on p.

PROOF. For $p = 2$ the proof can be obtained without special difficulty with the help of the Fourier transform and the Parseval equality (see [10] or [53]). In order to apply Theorem 1.5 and Corollary 1 it is necessary to verify that conditions (1.14) and (1.35) are satisfied. For this we use Corollary 2.

For the kernel $K(x,y) = \omega(x-y)/|x-y|^n$ we have

$$|\nabla_y K(x,y)| \leq \frac{C}{|x-y|^{n+1}}. \tag{1.42}$$

Expanding the domain of integration and applying (1.42), we get

$$h \int_{S'_{2h}} |\nabla_y K(x,y)|\, dx \leq Ch \int_{|x-y| \leq h/2} \frac{dx}{|x-y|^{n+1}} = 2C\sigma_{n+1}, \tag{1.43}$$

where σ_{n-1} is the "area" of the $(n-1)$-dimensional unit sphere. The same type of inequality holds also for the transposed kernel $K(x,y)$. Theorem 1.6 now follows from Corollary 2 and Theorem 1.5 for $q = 1$.

This theorem finds application in estimating the leading derivatives of solutions of elliptic boundary value problems. We present some simple examples.

LEMMA 1.1. *Suppose $u = u(x), x = (x_1, x_2, x_3)$, is a solution in the half-space $x_3 > 0$ of the boundary value problem*

$$\Delta u = \sum_{k=1}^{n} \frac{\partial g_k}{\partial x_k}; \tag{1.44}$$

$$u|_{x_3=0} = 0; \qquad u|_{|x| \to \infty} = 0, \tag{1.45}$$

where the g_k are smooth compactly supported functions. Then

$$\|D_x u\|_{L_p} \leq \frac{Cp^2}{p-1} \sum_{k=1}^{3} \|g_k\|_{L_p}, \tag{1.46}$$

where $p > 1$ and C is an absolute constant.

PROOF. We represent the solution of problem (1.44), (1.45) in the form

$$u = \frac{\partial u_1}{\partial x_1} + \frac{\partial u_2}{\partial x_2} + \frac{\partial u_3}{\partial x_3}, \tag{1.47}$$

where the functions u_1, u_2, and u_3 are determined by solving the boundary value problems

$$\Delta u_k = g_k; \qquad u_k|_{x_3=0} = 0 (k = 1, 2); \tag{1.48}$$

$$\Delta u_3 = g_3; \qquad \left.\frac{\partial u_3}{\partial x_3}\right|_{x_3=0} = 0; \tag{1.49}$$

$$u_k, u_3 \to 0 \qquad (|x| \to \infty). \tag{1.50}$$

From (1.47)–(1.50) it is evident that Lemma 1.1 follows immediately from the next lemma.

LEMMA 1.2. *Suppose a smooth function $u(x)$ vanishing at infinity in the half-space $x_3 > 0$ satisfies the Poisson equation*

$$\Delta u = g(x) \tag{1.51}$$

with a smooth compactly supported right side $g(x)$, and one of the boundary conditions

$$u|_{x_3=0} = 0; \tag{1.52}$$

$$\left.\frac{\partial u}{\partial x_3}\right|_{x_3=0} = 0. \tag{1.53}$$

Then

$$\|D_x^2 u\|_{L_p} \le C\frac{p^2}{p-1}\|g\|_{L_p}, \tag{1.54}$$

where C is an absolute constant.

PROOF. We extend the functions $u(x)$ and $g(x)$ to the half-space $x_3 < 0$ in an odd manner in the case of condition (1.52) and in an even manner in the case of (1.53). It then turns out that $u(x)$ satisfies the Poisson equation (1.51) in the entire space. As we know, the solution has the form

$$u(x) = -\frac{1}{4\pi}\int_{R^3}\frac{1}{|x-y|}g(y)\,dy. \tag{1.55}$$

Differentiating (1.55) according to familiar rules, we obtain

$$\frac{\partial^2 u}{\partial x_i \partial x_k} = \frac{1}{3}\delta_{ik}g(x) + \frac{1}{4\pi}\int_{R^3}|x-y|^{-3}\left[3\frac{(x_i-y_i)(x_k-y_k)}{|x-y|^2} - \delta_{ik}\right]g(y)\,dy. \tag{1.56}$$

It is easy to see that Theorem 1.6 is applicable to the singular integral in (1.56), and hence Lemma 1.2 and together with it Lemma 1.1 are proved.

Lemmas 1.2 and 1.1 extend to boundary value problems for arbitrary domains by means of the usual technique [86]–[89].

THEOREM 1.7. *In a bounded domain* Ω *with boundary* $S \in C^2$, *consider the boundary value problem for equation* (1.44) *with one of the conditions*

$$u/s = 0 \tag{1.57}$$

$$\left.\frac{\partial u}{\partial n}\right|_s = 0; \qquad \int_\Omega \sum_{k=1}^{n} \frac{\partial g_k}{\partial x_k}\, dx = 0. \tag{1.58}$$

Then

$$\|D_x u\|_{L_p} \le \frac{Cp^2}{p-1} \sum_{k=1}^{3} \|g_k\|_{L_p}. \tag{1.59}$$

3. Multipliers of series and Fourier integrals. The theorem of Marcinkiewicz on multipliers of Fourier series (see [33]) and the theorem of Mikhlin on multipliers of Fourier integrals (see [86]) together with estimates of singular integral operators are often used in investigating differential properties of solutions of elliptic and parabolic type. For convenience of reference we present the following theorem.

THEOREM 1.8. *Suppose the function* $\Phi(x)$ *is continuous for all* $x \in R^m$ *except possibly the point* 0. *Suppose the derivative* $\partial^m \Phi / \partial x_1 \cdots \partial x_m$ *exists, and all preceding derivatives are continuous. Suppose*

$$|x|^k \left| \frac{\partial^k \Phi}{\partial x_{j1} \partial x_{j2}, \dots, \partial x_{jk}} \right| \le M; 0 \le k \le m;$$
$$1 \le j_1 < j_2 < \cdots < j_k \le m.$$

Then the operator P, *given by*

$$(Pf)(x) = \frac{1}{(2\pi)^m} \int_{R^m} e^{i(x,y)} \Phi(y)\, dy \int_{R^m} e^{-i(y,z)} f(z)\, dz$$

is defined on a set dense in $L_p(R^m)$ *and is bounded in this space.*

§2. Some estimates of solutions of evolution equations

1. The Cauchy problem. Here we establish a number of estimates of the solution of the Cauchy problem for the evolution equation

$$\frac{du}{dt} + Au = f(t); \qquad u(0) = a, \tag{2.1}$$

and also estimates of periodic (or bounded) solutions. We suppose that A is a linear operator acting in a Banach space X which is possibly unbounded but is closed and has dense domain in X.

We suppose that the resolvent set of the operator A contains for some φ $(0 < \varphi < \pi/2)$ a sector $\Sigma_{\sigma_0,\varphi}$: $\varphi \le |\arg(\sigma - \sigma_0)| \le \pi$ (σ_0 is a real number) and for $\sigma \in \Sigma_{\sigma_0,\varphi}$ the resolvent $R_\sigma = R_\sigma(A) = (\sigma I - A)^{-1}$ satisfies the inequality

$$\|(\sigma I - A)^{-1}\|_{X \to X} \leq \frac{C}{|\sigma - \sigma_0| + 1}. \tag{2.2}$$

In deriving sharp estimates a more special and stronger assumption will be used. Namely, we shall assume that (2.2) is satisfied not for the single space X but for any of the spaces $L_p(\Omega, \mu, E)$ ($p > 1$; E is a Banach space)

$$\|(\sigma I - A)^{-1}\|_{L_p \to L_p} \leq \frac{C_p}{|\sigma - \sigma_0| + 1}. \tag{2.3}$$

Here the constant C_p and the angle φ can depend on p.

As is known [22], [74], conditions (2.2) are sufficient in order that the operator $-A$ generate a strongly continuous semigroup $U(t) = e^{-tA}$ analytic in t in some sector. Condition (2.3) was derived by Solomyak [74] in the case where A is an elliptic operator of order $2m$ under the conditions of the first boundary value problem, while for general boundary conditions this was done by Agmon and Nirenberg [2]. It is shown below in §6 that the linearized Navier-Stokes operator satisfies this condition. This result was formulated in [90]. It was noted in [72] that condition (2.3) is a consequence of the estimates of the leading derivatives of solutions of the time-dependent Navier-Stokes equations given in [84], [78], and [77]. A direct proof is presented in §6.

We introduce the following definition. Suppose a closed linear operator $L: X \to Y$ (Y is a Banach space) is defined on the range of the operator R_σ for any $\sigma \in \Sigma_{\sigma_0,\varphi}$, and that

$$\|LR_\sigma\|_{X \to Y} \leq \frac{C}{|\sigma|^{1-\alpha}} \qquad (0 \leq \alpha \leq 1). \tag{2.4}$$

We then say that the operator L has *degree* α *relative to the operator* A. It is possible to give another definition which is also valid for $\alpha > 1$. Namely, we say that L has degree $\alpha > 0$ relative to the operator A if Le^{-tA} is a bounded operator for any $t > 0$ and

$$\|Le^{-tA}\|_{X \to Y} \leq \frac{C(t)}{t^\alpha},$$

where $C(t)$ is a continuous function of t.

In particular, as we know,

$$\|e^{-tA}\|_{X \to X} \leq Ce^{-\sigma_0 t};$$

$$\|Ae^{-tA}\|_{X \to X} \leq \frac{C}{t} e^{-\sigma_0 t}.$$

Therefore, the operator A has degree 1 relative to itself, while the identity operator $I: X \to X$ has degree 0 relative to A.

LEMMA 2.1. *Suppose the operator L has degree α relative to the operator A. Then for $t > 0$*

$$\|Le^{-tA}\|_{X\to Y} \le \frac{C}{t^\alpha} e^{-\sigma_0 t}. \tag{2.5}$$

PROOF. For the semigroup e^{-tA} under condition (2.2) there is the integral representation [22], [37]

$$e^{-tA} = \frac{1}{2\pi i} \int_\gamma e^{-\lambda t} R_\lambda \, d\lambda, \tag{2.6}$$

where γ is the boundary of the sector $\Sigma_{\sigma_0,\varphi}$. Applying the operator L to (2.6) and using (2.4), we find

$$\begin{aligned} \|Le^{-tA}\|_{X\to Y} &\le \frac{1}{2\pi} \int_\gamma |e^{-\lambda t}| \cdot \|LR_\lambda\| \cdot |d\lambda| \\ &\le C \int_0^\infty e^{-(\sigma_0 + r\cos\varphi)t} \frac{dr}{r^{1-\alpha}} \\ &= Ce^{-\sigma_0 t} \cdot \frac{1}{t^\alpha} \int_0^\infty e^{-r\cos\varphi} \frac{dr}{r^{1-\alpha}}, \end{aligned} \tag{2.7}$$

which essentially coincides with (2.5). The lemma is proved.

The solution of the Cauchy problem (2.1) can be written in the form

$$u(t) = e^{-tA} a + \int_0^t e^{-(t-\tau)A} f(\tau) \, d\tau. \tag{2.8}$$

We introduce the notation

$$U_1 a = u_1(t) = e^{-tA} a; u_2(t) = \int_0^t e^{-(t-\tau)A} f(\tau) \, d\tau = U_2 f. \tag{2.9}$$

Below, a number of estimates for the operators U_1 and U_2 are given which show in what sense the (so far formal) solution (2.9) is to be understood.

LEMMA 2.2. *Suppose the operator $L\colon X \to Y$ has degree α $(0 < \alpha < 1)$ relative to the operator A. Then*

$$\left(\int_0^T \|Lu_2(t)\|_Y^q e^{q\sigma_0 t} \, dt\right)^{1/q} \le C \left[\int_0^T \|f(\tau)\|_X^p e^{-p\sigma_0 t} \, d\tau\right]^{1/p};$$
$$\left(1 \le p < \frac{1}{1-\alpha}; \qquad q = \frac{p}{1-(1-\alpha)p}\right). \tag{2.10}$$

Here T is any positive number (or $+\infty$); C does not depend on T.

PROOF. Using Lemma 2.1, we obtain

$$\begin{aligned}\|Lu_2(t)\|_Y &= \left\|\int_0^t Le^{-(t-\tau)A} f(\tau)\,d\tau\right\|_Y \\ &\le \int_0^t \|Le^{-(t-\tau)A}\|_{X\to Y}\cdot\|f(\tau)\|_X\,d\tau \\ &\le \int_0^t \frac{C}{(t-\tau)^\alpha} e^{-\sigma_0(t-\tau)}\|f(\tau)\|_X\,d\tau.\end{aligned} \tag{2.11}$$

According to the theorem on integrals of potential type, (2.10) now follows from (2.11). The lemma is proved.

LEMMA 2.3. *If $p > 1/(1-\alpha), 0 \le \alpha < 1$, then $Lu_2(t)$ satisfies a Hölder condition with exponent $\beta = 1/p' - \alpha$; more precisely,*

$$\sup_{0\le t\le T}\sup_{h>0} e^{\sigma_0' t}\frac{|Lu_2(t+h) - Lu_2(t)|_Y}{h^\beta} \le C\left[\int_0^T |f(\tau)|_X^p e^{p\sigma_0\tau}\,d\tau\right]^{1/p}. \tag{2.12}$$

PROOF. For simplicity we set $\sigma_0 = 0$ (we can arrive at this case by the change $A \to A - \sigma_0 I$). We proceed from the relation

$$\begin{aligned}Lu_2(t+h) - Lu_2(t) = &\int_0^{t+h} Le^{-(t+h-\tau)A} f(\tau)\,d\tau \\ &- \int_0^t Le^{-(t-\tau)A} f(\tau)\,d\tau.\end{aligned} \tag{2.13}$$

From (2.13) we obtain

$$\begin{aligned}|Lu_2(t+h) - Lu_2(t)|_Y \le &\left|\int_0^{t-h} L(e^{-(t+h-\tau)A} - e^{-(t-\tau)A}) f(\tau)\,d\tau\right|_Y \\ &+ \left|\int_{t-h}^{t+h} Le^{-(t+h-\tau)A} f(\tau)\,d\tau\right|_Y \\ &+ \left|\int_{t-h}^t Le^{-(t-\tau)A} f(\tau)\,d\tau\right|_Y \equiv I_1 + I_2 + I_3.\end{aligned} \tag{2.14}$$

We represent the operator contained in the integrals I_1 and I_2 in the form of a product

$$L(e^{-(t+h-\tau)A} - e^{-(t-\tau)A}) = (Le^{-1/2(t-\tau)A})\cdot(Ae^{-1/2(t-\tau)A})\left(\frac{e^{-hA} - I}{A}\right). \tag{2.15}$$

We estimate the first two factors in (2.15) with the help of inequalities of the type (2.5) with consideration of the known fact that A has degree 1 relative to itself. To estimate the last factor in (2.15) we use the equation which the semigroup e^{-tA} satisfies

$$\frac{d}{dt}e^{-tA} = -Ae^{-tA}. \tag{2.16}$$

Integrating (2.16) from 0 to h, we obtain

$$A^{-1}(e^{-hA} - I) = -\int_0^h e^{-tA}\,dt. \tag{2.17}$$

From this it follows that

$$\|A^{-1}(e^{-hA} - I)\|_{X\to X} \leq Ch. \tag{2.18}$$

Thus,

$$\|L(e^{-(t+h-\tau)A} - e^{-(t-\tau)A})\|_{X\to Y} \leq \frac{Ch}{(t-\tau)^{\alpha+1}}. \tag{2.19}$$

Further, we have

$$I_1 \leq Ch\int_0^{t-h} \frac{|f(\tau)|_X}{(t-\tau)^{\alpha+1}}\,d\tau, \tag{2.20}$$

whence by applying the Hölder inequality we find that

$$I_1 \leq Ch\left(\int_0^T |f(\tau)|_X^p\,d\tau\right)^{1/p}\cdot\left(\int_{-\infty}^{t-h}\frac{d\tau}{(t-\tau)^{(\alpha+I)_{p'}}}\right)^{1/p'} = C_1h^\beta\|f\|_{L_p}; \tag{2.21}$$

$$C_1 = C[(\alpha+1)p'-1]^{1/p'}; \qquad \frac{1}{p}+\frac{1}{p'} = 1.$$

Applying (2.5), for the integral I_2 we obtain

$$\begin{aligned} I_2 &\leq C\int_{t-h}^{t+h}\frac{|f(\tau)|_X}{(t+h-\tau)^\alpha}\,d\tau \\ &\leq C\|f\|_{L_p}\left[\int_{t-h}^{t+h}\frac{d\tau}{(t+h-\tau)^{\alpha p'}}\right]^{1/p'} \\ &= C_2h^\beta\|f\|_{L_p}; \\ &\quad C_2 = C\cdot 2^{1/p'-\alpha}(1-\alpha p')^{-1/p'}. \end{aligned} \tag{2.22}$$

Similarly, for I_3,

$$\begin{aligned} I_3 \leq C\int_{t-h}^{t}\frac{|f(\tau)|_X}{(t-\tau)^\alpha}\,d\tau &\leq C\|f\|_{L_p}\cdot\left[\int_{t-h}^{t}\frac{d\tau}{(t-\tau)^{\alpha p'}}\right]^{1/p'} \\ &= C_3h^\beta\|f\|_{L_p}; \\ &\quad C_3 = C(1-\alpha p')^{-1/p'}. \end{aligned} \tag{2.23}$$

The estimate (2.12) follows immediately from (2.14) and (2.21)–(23). The lemma is proved.

We further consider the limit case $p = 1/(1-\alpha)$. In this case in (2.10) q may be assumed arbitrary, and

$$\left(\int_0^T \|Lu_2(t)\|_Y^q e^{-q\sigma_0 t}\,dt\right)^{1/q} \le Cq^\alpha \left[\int_0^T \|f(t)\|_X^p e^{-p\sigma_0 t}\,dt\right]^{1/p}, \quad (2.24)$$

where C does not depend on $q, q > 1$, and $\sigma_0 < \sigma$. It follows from (2.24) that the following integral is bounded for any $T > 0$:

$$\int_0^T \exp(\gamma\|Lu_2(t)\|_Y^{1/\alpha})\,dt < \infty; \qquad \gamma < \alpha C^{-1/\alpha}e^{-1} \quad (2.25)$$

LEMMA 2.4. *Suppose the operator L has degree α $(0 < \alpha < 1)$ relative to A. Then*

$$\left(\int_0^\infty |Lu_1(\tau)|_Y^q \tau^\varepsilon e^{-q\sigma_0\tau}\,d\tau\right)^{1/q} \le C_\varepsilon\|a\|_X;$$

$$q \le q_0 = \frac{1}{\alpha}, \quad (2.26)$$

where $\varepsilon > 0$ is any number, and C_ε depends on ε but not on a; for $q < q_0$ it may be assumed that $\varepsilon = 0$.

PROOF. We observe that if $\delta > 0$ is so small that the spectrum of A still lies to the right of the line $\operatorname{Re}\sigma = \sigma_0 + \delta$, then in (2.5) in place of σ_0 it is possible to take $\sigma_0 + \delta$. Therefore

$$|Lu_1(\tau)|_Y = |Le^{-\tau A}a|_Y \le \frac{C}{\tau^\alpha}e^{-1|\sigma_0+\delta|\tau}\|a\|_X. \quad (2.27)$$

Inequality (2.26) follows immediately from (2.27). Moreover,

$$C_\varepsilon = C\int_0^\infty e^{-\delta q\tau}\tau^{\varepsilon-\alpha q}\,d\tau = C(\delta q)^{\alpha q-\varepsilon-1}\Gamma(1-\alpha q+\varepsilon). \quad (2.28)$$

It is not known whether in the general case we can take $\varepsilon = 0$ in this lemma. It is essential, however that we be able to do this when condition (2.3) is satisfied.

LEMMA 2.5. *Suppose the operator $A\colon L_p(\Omega,\mu,E) \to L_p(\Omega,\mu,E)$ satisfies condition (2.3) for any $p > 1$. Suppose the operator $L\colon L_p(\Omega,\mu,E) \to Y$ for any $p > 1$ has relative to A fractional degree $\alpha_p = \beta/p + \gamma$ (β and γ do not depend on p, and $0 < p\alpha_p \le 1$ for $1 \le p_1 < p < p_2$). Then for any $p \in (p_1, p_2)$*

$$\left(\int_0^\infty |Lu_1(\tau)|_Y^{q\sigma_0\tau}\,d\tau\right)^{1/q} \le C\|a\|_{L_p(\Omega,\mu,E)}, \quad (2.29)$$

where $q = 1/\alpha_q$; c does not depend on a.

PROOF. We shall treat $R = Le^{-tA}$ as an operator from $L_p(\Omega, \mu, E)$ to $L_q((0,\infty), m_1, Y)$, and show that the conditions of Marcinkiewicz interpolation theorem are satisfied. Applying Lemma 2.1, we obtain

$$|Lu_1(\tau)|_Y = |Le^{-tA}a|_Y \leq \frac{Ce^{-\sigma_0,t}}{t^{\alpha p}}\|a\|_{L_p(\Omega,\mu,E)} \tag{2.30}$$

It remains to apply Theorem 1.3, and the lemma is proved.

LEMMA 2.6. *Suppose the operator* $L\colon X \to Y$ *has degree* α $(1 < \alpha < 1)$ *relative to* A. *Then*

$$\|Lu_2(t)\|_Y \leq Ce^{-\sigma_0 t}t^{1/r'-\alpha}\left(\int_0^t e^{r\sigma_0\tau}\|f(\tau)\|_X^r\,d\tau\right)^{1/r}, \tag{2.31}$$

where $r > 1/(1-\alpha)$.

PROOF. Applying Lemma 2.1 and the Hölder inequality, we obtain

$$\begin{aligned}\|Lu_2(t)\|_y &\leq C\int_0^t \frac{e^{-\sigma_0(t-\tau)}}{(t-\tau)^\alpha}\|f(\tau)\|_X\,d\tau \\ &\leq Ce^{-\sigma_0 t}\left(\int_0^t e^{r\sigma_0\tau}\|f(\tau)\|_X^r\,d\tau\right)^{1/r}\cdot\left[\int_0^t \frac{d\tau}{(t-\tau)^{\alpha r'}}\right]^{1/r'},\end{aligned} \tag{2.32}$$

which coincides with (2.31). The lemma is proved.

Again, in the case where condition (2.3) is satisfied it is possible to obtain a sharper result.

LEMMA 2.7. *Suppose the operator* $A\colon L_p(\Omega, \mu, E) \to L_p(\Omega, \mu, E)$ *satisfies condition* (2.3) *for any* $p > 1$. *Suppose the operator* $L\colon L_p(\Omega, \mu, E) \to L_{m_p}(\Omega_1, \mu_1, E_1)$ *has degree* α_p $(0 < \alpha_p < 1)$ *relative to* A *for any* p *in some interval* (p_0, p_1), *where* $\alpha_p = \beta/p + \gamma$ *and* $1/m_p = \beta_1/p + \gamma_1 \leq 1 - \alpha_p$ $(p_0 < p < p_1)$, *while the constants* α, β, α_1, *and* β_1 *do not depend on* p. *Then*

$$\|Lu_2(t)\|_{L_{m_p}(\Omega_1,\mu_1,E_1)} \leq Ce^{-\sigma_0 t}\left(\int_0^t e^{r_p\sigma_0\tau}\|f(\tau)\|_{L_p}^{r_p}\,d\tau\right)^{1/r};\quad \left(r_p = \frac{1}{1-\alpha_p}\right). \tag{2.33}$$

PROOF. In order to simplify notation we write L_p and L_{m_p} in place of $L_p(\Omega, \mu, E)$ and $L_{m_p}(\Omega_1, \mu_1, E_1)$. As previously, we may take $\sigma_0 = 0$ without losing anything.

We consider the operator $K\colon L_{r_p}((0,t), m_1, L_p) \to L_{m_p}$ defined for fixed $t > 0$ by

$$g = Kf = Lu_2 = \int_0^t Le^{-(t-\tau)A} f(\tau)\, d\tau. \tag{2.34}$$

The problem consists in proving that it is bounded. In place of this it suffices to prove that the adjoint operator K^* is bounded. As is known [22], any bounded linear functional g_* on $L_{m_p}(\Omega_1, \mu_1, E_1)$ can be realized in the form

$$(g_*, g) = \int_{\Omega_1} (g_*^0(\omega_1), g(\omega_1))\, d\mu_1(\omega_1), \tag{2.35}$$

where the function $g_*^0 \in L_{m_p'}(\Omega_1, \mu_1, E^*)$ is uniquely defined by the functional, and the mapping $g_* \to g_*^0$ is isometric. In this sense $L_{m_p}^*(\Omega_1, \mu_1, E_1) = L_{m_p'}(\Omega_1, \mu_1, E^*)$. In exactly the same way, $L_{r_p}^*((0,t), m_1, L_p) = L_{r_p'}((0,t), m_1, L_p^*)$.

We consider the identity

$$(g_*, Kf) = \int_0^t (e^{-(t-\tau)A^*} L^* g_*, f(r))\, d\tau. \tag{2.36}$$

Here we have used the well-known fact, which is easily derived from (2.6), that

$$(e^{-(t-\tau)A})^* = e^{-(t-\tau)A^*}. \tag{2.37}$$

From (2.36) it follows that the operator $K^*\colon L_{m_p'} \to L_{r_p}'((0,t), m_1, L_p^*)$ has the form

$$(K^* g_*)(\tau) = e^{-(t-\tau)A^*} L^* g_*. \tag{2.38}$$

Applying Lemma 2.1, for fixed $\tau < t$ we obtain

$$\|e^{-(t-\tau)A^*} L^*\|_{L_{m_p'} \to L_{p'}} = \|Le^{-(t-\tau)A}\|_{L_p \to L_{m_p}} \le \frac{C}{(t-\tau)^{\alpha_p}}. \tag{2.39}$$

Therefore, from (2.38) we get

$$|(K^* g_*)(\tau)|_{L_{p'}} \le \frac{C}{(t-\tau)^{\alpha_p}} \|g_*\|_{L_{m_p'}}. \tag{2.40}$$

The boundedness of the operator K^* and hence of K follows from (2.40) according to the Marcinkiewicz interpolation theorem (see Theorem 1.3). The lemma is proved.

2. Bounded, periodic, and almost periodic solutions. We now take up bounded solutions of the differential equation

$$\frac{du}{dt} + Au = f(t). \tag{2.41}$$

We additionally assume that the spectrum of the operator A does not contain points of the imaginary axis:

$$\operatorname{Im} \sigma(A) \neq 0. \tag{2.42}$$

Let $\sigma_+ = \sigma_+(A)$ and $\sigma_- = \sigma_-(A)$ be the parts of the spectrum of A in the right and left half-planes, respectively. We introduce the corresponding projections

$$P_- = \frac{1}{2\pi i} \int_{\gamma_-} R_\lambda \, d\lambda; \qquad P_+ = I - P_-, \tag{2.43}$$

where γ_- is a smooth contour lying entirely in a bounded portion of the left half-plane and surrounding the set σ_-. The operator A can be naturally represented as the sum

$$A = A_+ + A_-; \qquad A_+ = P_+ A; A_- = P_- A. \tag{2.44}$$

The space X is thus decomposed into a direct sum of subspaces $X_+ = P_+(X)$ and $X_- = P_-(X)$. The subspaces X_+ and X_- are invariant relative to the operator A: they contain dense linear manifolds $P_+(D_A)$ and $P_-(D_A)$ on which A is defined and which A takes into X_+ and X_-, respectively.

LEMMA 2.8. *The projections P_+ and P_- act boundedly in X; the operator A_- admits continuation to a bounded operator; and the operator $-A_+$ is the generator of an analytic semigroup on X_+ and on X. Here*

$$P_+ A u = A P_+ u; \qquad P_- A u = A P_- u \qquad (u \in D_A); \tag{2.45}$$

$$U(t) \equiv e^{-tA} = U_+(t) + U_-(t); \qquad U_\mp(t) = e^{-tA_\mp} P_\mp. \tag{2.46}$$

PROOF. Boundedness of P_- follows immediately from (2.43), since the contour γ_- lies in a bounded portion of the plane and the resolvent R_λ is analytic on it. By (2.43) P_+ is then also bounded. The fact that A commutes with P_+ and P_- is well known and follows immediately from (2.43) and the fact that A commutes with the resolvent R_λ.

Further, for the operator A_- there is the representation

$$A_- u = \frac{1}{2\pi i} \int_{\gamma_-} \lambda R_\lambda u \, d\lambda, \tag{2.47}$$

which for $u \in D_A$ follows immediately from (2.43) and the identity

$$A R_\lambda = -I + \lambda R_\lambda \tag{2.48}$$

with the use of Cauchy's integral theorem. Boundedness of A_- follows immediately from (2.47).

We now consider the operator A_+. Its spectrum is the set $\sigma_+(A) \cup \{0\}$, while the resolvent is

$$(\lambda I - A_+)^{-1} = P_+(\lambda I - A)^{-1} + \frac{1}{\lambda} P_-. \tag{2.49}$$

Because of (2.49) and (2.2) the operator A_+ satisfies condition (2.2) and hence generates an analytic semigroup on X and so also on the invariant subspace X_+. The latter can be written in the form

$$e^{-tA_+} P_+ = P_+ e^{-tA_+}. \tag{2.50}$$

We deduce the representation (2.46) directly from (2.6) by breaking the integral over γ into a sum of integrals over γ_- and γ_+. The lemma is proved.

THEOREM 2.1. *Suppose the spectrum of the operator A does not contain points of the imaginary axis, $f(t)$ is a measurable function, and the quantity $\|f(t)\|_X$ is bounded for $t \in (-\infty, +\infty)$. Equation* (2.41) *then has a unique bounded solution*

$$u_0(t) = \int_{-\infty}^{t} e^{-(t-\tau)A_+} P_+ f(\tau)\, d\tau - \int_t^{\infty} e^{-(t-\tau)A} P_- f(\tau)\, d\tau. \tag{2.51}$$

If $L\colon X \to Y$ is an operator of degree α $(0 \le \alpha < 1)$ relative to A, then

$$\sup_{-\infty \le t < \infty} \|Lu_0(t)\|_Y \le C \sup_{-\infty < t < \infty} \|f(t)\|_X. \tag{2.52}$$

PROOF. We begin with the proof of uniqueness. For this we establish that the homogeneous equation

$$\frac{du}{dt} + Au = 0 \tag{2.53}$$

has no nonzero bounded solutions. Any solution of (2.53) can be represented in the form

$$u(t) = u_+(t) + u_-(t); \qquad u_\pm(t) = P_\pm u(t) = e^{-(t-\tau)A_\mp} u_\mp(\tau), \tag{2.54}$$

where $t > \tau$. According to the hypothesis of the theorem, there exists $\sigma_1 > 0$ such that in the strip $|\operatorname{Re}\sigma| \le \sigma_1$ there are no points of the spectrum of A. We therefore have

$$\begin{aligned} \|u_+(t)\| &\le Ce^{-\sigma_1(t-\tau)}\|u_+(\tau)\|; \\ \|u_-(t)\| &\le Ce^{\sigma_1(t-\tau)}\|u_-(\tau)\|. \end{aligned} \tag{2.55}$$

It follows from (2.55) that the vector-valued function $u_+(t)$ is bounded for $t \ge \tau$. By (2.54) and the boundedness of $u(t)$ it then follows that $u_-(t)$ is bounded for $t \geq \tau$. A similar argument shows that boundedness also holds for $t < \tau$, and hence $u_+(t)$ and $u_-(t)$ are bounded. From (2.55) we obtain

$$\|u_+(\tau)\| \geq \frac{1}{C} e^{\sigma_1(t-\tau)} \|u_+(t)\|;$$
$$\|u_-(\tau)\| \geq \frac{1}{C} e^{-\sigma_1(t-\tau)} \|u_-(t)\|. \tag{2.56}$$

If it is assumed, for example, that $u_+(t) \neq 0$, from (2.56) it follows that $\underline{\lim}_{\tau \to -\infty} \|u_+(\tau)\| = +\infty$. Hence $u_+(t) \equiv 0$, and similarly $u_-(t) \equiv 0$, which proves uniqueness.

We proceed to the derivation of (2.52). Using Lemma 2.1, we get

$$\begin{aligned} \|Lu_0(t)\|_y &\leq C_1 \left[\int_{-\infty}^{t} \frac{e^{-\sigma_1(t-\tau)}}{(t-\tau)^\alpha} d\tau + \int_t^\infty \frac{e^{-\sigma_0(\tau-t)}}{(\tau-t)^\alpha} d\tau \right] \times \sup_t \|f(t)\|_X \\ &= C_1 \left[\int_0^\infty \frac{e^{-\sigma_1 s}}{s^\alpha} ds + \int_0^\infty \frac{e^{-\sigma_0 s}}{s^\alpha} ds \right] \sup_t \|f(t)\|_X \\ &= C \sup_t \|f(t)\|_X. \end{aligned} \tag{2.57}$$

The theorem is proved.

We remark that if the spectrum of A contains points of the imaginary axis, then for some bounded right sides equation (2.41) will not have a bounded solution. This is proved in essentially the same way as in [38].

Existence of a bounded solution can be proved under fewer conditions on the right side f. Let R be a Banach space of functions defined on the segment $[0, T]$ $(T > 0)$ and having range in X. We denote by SR the space of functions $f(t)$ defined on the real axis with values in X and such that $f(t+a) = f_a(t)$ for any a coincides on $[0, T]$ with some function in R. We define the norm in SR by

$$\|f\|_{SR} = \sup_{-\infty < a < +\infty} \|f_a\|_R. \tag{2.58}$$

In particular, if $R = L_p((0, T), m_1, X)$ $(p \geq 1)$, we have

$$\|f\|_{SL_p} = \sup_a \left(\int_0^T |f(t+a)|_X^p \, dt \right)^{1/p}. \tag{2.59}$$

The closure of the set of trigonometric quasipolynomials with coefficients in X in the norm (2.59) is a vector-valued generalization of Stepanov's almost periodic functions.

THEOREM 2.2. *Suppose that the operators A and L are the same as in Theorem 2.1, $f \in SL_p((0, T), m_1, X)$, and $1 - \alpha p' > 0$. Then equation*

(2.41) *has the unique bounded solution* (2.51), *and*

$$\sup_t |Lu_0(t)| \le C\|f\|_{SL_p};$$
$$\sup_{t,\tau} \frac{|Lu_0(t) - Lu_0(\tau)|}{|t-\tau|^\beta} \le C\|f\|_{SL_p}; \qquad \beta = \frac{1}{p'} - \alpha. \tag{2.60}$$

If $1 - \alpha p' < 0$, *then instead of* (2.60)

$$\|Lu_0\|_{SL_r} \le C\|f\|_{SL_p}; \qquad r = \frac{p}{1-(1-\alpha)p}. \tag{2.61}$$

PROOF. To simplify notation we assume that the spectrum of A lies in the half-plane $\operatorname{Re}\sigma \ge \sigma_0 > 0$. In (2.51) only the first term then remains, and

$$u_0(t) = \int_{-\infty}^{t} e^{-(t-\tau)A} f(\tau)\, d\tau = \sum_{k=0}^{\infty} \int_{t-(k+1)T}^{t-kT} e^{-(t-\tau)A} f(\tau)\, d\tau. \tag{2.62}$$

Applying Lemma 2.1, from (2.62) we deduce that

$$\begin{aligned} |Lu_0(t)|_Y &\le C_1 \sum_{k=0}^{\infty} \int_{t-(k+1)T}^{t-kT} \frac{e^{-\sigma_0(t-\tau)}}{(t-\tau)^\alpha} |f(\tau)|_X\, d\tau \\ &\le C_1 \sum_{k=0}^{\infty} e^{-\sigma_0 kT} \int_{t-(k+1)T}^{t-kT} \frac{|f(\tau)|_X}{(t-\tau)^\alpha}\, d\tau. \end{aligned} \tag{2.63}$$

In the case $1 - \alpha p' > 0$, estimating the integrals in (2.63) by the Hölder inequality, we obtain

$$\begin{aligned} |Lu_0(t)_Y| &\le C_1 \|f\|_{SL_p} \cdot \sum_{k=0}^{\infty} e^{-\sigma_0 kT} \left[\int_{t-(k+1)T}^{t-kT} \frac{d\tau}{(t-\tau)^{\alpha p'}} \right]^{1/p'} \\ &\le C_1 (1-\alpha p')^{-1/p'} T^{(1-\alpha p')/p'} \|f\|_{SL_p} \\ &\quad \cdot \sum_{k=0}^{\infty} e^{-\sigma_0 kT} [(k+1)^{1-\alpha p'} - k^{1-\alpha p'}]^{1/p'} \\ &= C\|f\|_{SL_p}. \end{aligned} \tag{2.64}$$

This coincides with the first inequality of (2.60). The second inequality of (2.60) can be proved just as Lemma 2.3, and we hence omit the proof. Finally, applying the theorem on potentials to estimate the integrals on the right side of (2.63), we arrive at (2.61). The theorem is proved.

THEOREM 2.4. *Suppose the operators A and L are the same as in Theorem 2.3. If $f(t)$ is periodic in t with period T, then equation* (2.41) *has a* *unique periodic solution* (2.51) *with the same period* T. *If* $f(t)$ *is almost periodic in the Bohr or Smirnov sense, then the solution* $u_0(t)$ *and also* $Lu_0(t)$ *have this same property.*

PROOF. Periodicity of the solution (2.51) for a periodic right side f is obvious. Further, if f is a quasipolynomial,

$$f(t) = \sum_{k=-n}^{n} e^{i\lambda_k t} f_k; \qquad f_k \in X, \tag{2.65}$$

then so is the corresponding bounded solution of (2.41):

$$u_0(t) = \sum_{k=-n}^{n} e^{i\lambda_k t}(i\lambda_k I + A)^{-1} f_k. \tag{2.66}$$

We recall that the almost periodic functions of Stepanov and Bohr are obtained as a result of closing the set of quasipolynomials in the metric of SL_p and in the uniform metric respectively. Our assertions therefore follow from the preceding theorem (even with some strengthening).

If the spectrum of the operator A is decomposed into nonintersecting closed components $\sigma_+(A), \sigma_-(A)$, and $\sigma_0(A)$ lying respectively in the right half-plane, the left half-plane, and on the imaginary axis, then existence and uniqueness of a T-periodic solution hold if $\sigma_0(A)$ does not contain the points $2k\pi i/T$ $(k = 0, \mp 1, \dots)$.

§3. Estimates of the "leading derivatives" of solutions of evolution equations

In this section we continue the study of the Cauchy problem

$$\frac{du}{dt} + Au = f(t); \qquad u(0) = a, \tag{3.1}$$

and also of the corresponding periodic problem in the case where $f \in L_p$. Regarding the element a, we suppose that there exists a vector-valued function $v(t)$ such that $v(0) = a$ and

$$\left\| \frac{dv}{dt} \right\|_{L_p((0,T),X)} + \|Av\|_{L_p((0,T),X)} < \infty; \tag{3.2}$$

here $T > 0$ is some number.

Such vectors a have the finite seminorm (it often turns out to be a norm)

$$|||a|||_p = \inf_v \left(\left\| \frac{dv}{dt} \right\|_p^p + \|Av\|_p^p \right)^{1/p}. \tag{3.3}$$

The infimum here is taken over all v satisfying (3.2). We shall be interested in conditions under which

$$\left[\int_0^T \left(\left|\frac{du(t)}{dt}\right|_X^p + |Au(t)|_X^p\right) dt\right]^{1/p} \leq C(\|f\|_{L_p((0,T),X)} + |||a|||_p) \qquad (3.4)$$

or in the case of a T-periodic right side f the analogous estimate of the periodic solution u_0 holds:

$$\left(\int_0^T \left(\left|\frac{du_0}{dt}\right|_X^p + |Au_0|_X^p\right) dt\right)^{1/p} \leq C\|f\|_{L_p((0,T),X)}. \qquad (3.5)$$

In the second part of this section estimates of du/dt and Au in Hölder norms are also considered.

1. Estimates in L_p. We remark that it suffices to obtain the estimate (3.4) for $a = 0$. Indeed, we can go over to this case by means of the change

$$u(t) = u_1(t) + v(t), \qquad (3.6)$$

where v is a vector-valued function satisfying (3.2). For u_1 we then obtain the problem

$$\frac{du_1}{dt} + Au_1 = f - \frac{dv}{dt} - Av; \qquad u_1(0) = 0. \qquad (3.7)$$

If the estimate (3.4) for $a = 0$ has already been proved, then from (3.6) and (3.7) we obtain

$$\left\|\frac{du}{dt}\right\|_p + \|Au\|_p \leq \left\|\frac{dv}{dt}\right\|_p + \|Av\|_p + C\left(\|f\|_p + \left\|\frac{dv}{dt}\right\|_p + \|Av\|_p\right). \qquad (3.8)$$

Passing to the infimum over all v satisfying conditions (3.2) in (3.8), we arrive at (3.4).

THEOREM 3.1. *Suppose the vector A satisfies condition* (2.2) *and the estimate* (3.4) *holds for some $p_0 > 1$. It is then satisfied for any $p > 1$. Moreover, for the solution of problem* (3.1) *for $a = 0$ the estimate*

$$\left[\int_0^T e^{p\sigma_0 t}\left(\left|\frac{du(t)}{dt}\right|_X^p + |Au(t)|_X^p\right) dt\right]^{1/p} \leq C_p\left[\int_0^T e^{p\sigma_0 t}|f(t)|_X^p\, dt\right]^{1/p}, \qquad (3.9)$$

holds, where $C_p = Cp^2/(p-1)$ and the constant C does not depend on f, T, or p.

PROOF. As shown above, it may be assumed that $a = 0$. In this case the solution of problem (3.1) has the form

$$u(t) = \int_0^t e^{-(t-\tau)A} f(\tau)\, d\tau.$$

For a set M of functions $f(t)$ continuous together with $Af(t)$ which is dense in $L_p(0, T)$ we can write

$$Au(t) = \int_0^t e^{-(t-\tau)A} Af(\tau)\, d\tau \equiv (Qf)(t). \tag{3.10}$$

For any $f \in L_p$, as before, we write (3.10), understanding the integral as a singular integral. Here there are various possiblities. We shall use the following definition. For $\varepsilon > 0$ we set

$$(Q_\varepsilon f)(t) = (e^{-\varepsilon A} Qf)(t) = \int_0^t Ae^{-(t-\tau+\varepsilon)A} f(\tau)\, d\tau. \tag{3.11}$$

We define the operator Q for any $f \in L_p(0, T)$ by setting

$$Qf = \lim_{\varepsilon \to 0} Q_\varepsilon f. \tag{3.12}$$

For $f \in M$ this definition obviously coincides with (3.10). By the hypothesis of the theorem the operator Q acts boundedly in L_{p_0}. From the definition (3.11) it follows that Q also possesses the same property, and its norm in L_{p_0} is bounded uniformly with respect to $\varepsilon > 0$:

$$\|Q_\varepsilon\|_{L_p \to L_{p_0}} \le \|e^{-\varepsilon A}\|_{X \to X} \cdot \|Q\|_{L_{p_0} \to L_{p_0}} \le C \|Q\|_{L_{p_0} \to L_{p_0}}. \tag{3.13}$$

We now use the extrapolation Theorem 1.5. As before, for brevity we can set $\sigma_0 = 0$. We define the kernel K_ε by setting

$$K_\varepsilon(t, \tau) = \begin{cases} Ae^{-(t-\tau+\varepsilon)A} & 0 \le \tau \le t \le T; \\ 0 & T \ge \tau > t \ge 0. \end{cases} \tag{3.14}$$

We shall verify that condition (1.14) for $q = 1$ holds for the kernel K_ε and the estimate is uniform with respect to ε. Let S_h denote the segment $[t_0 - h, t_0 + h]$, $0 \le t_0 \le T$, $h > 0$. It is necessary to obtain the estimate

$$I = \int_{S'_{2h} \cap [0,T]} |K_\varepsilon(t, \tau) - K_\varepsilon(t, \tau')|\, dt \le B, \tag{3.15}$$

where $\tau, \tau' \in S_h$, and the constant B does not depend on t_0, τ, τ', h, or ε. We first assume that $0 \leq t_0 - 2h < t_0 + 2h \leq T$. We then obtain

$$I = \int_0^{t_0-2h} \|K_\varepsilon(t,\tau) - K_\varepsilon(t,\tau')\|\,dt + \int_{t_0+2h}^{T} \|K_\varepsilon(t,\tau) - K_\varepsilon(t,\tau')\|\,dt. \tag{3.16}$$

Considering (3.14), we reduce (3.16) to the form

$$I = \int_{t_0+2h}^{T} \|K_\varepsilon(t,\tau) - K_\varepsilon(t,\tau')\|\,dt. \tag{3.17}$$

We shall estimate the integrand in (3.17). With consideration of (2.2), (2.60), and the theorem on finite increments we have

$$\begin{aligned}|K_\varepsilon(t,\tau) - K_\varepsilon(t,\tau')| &= \frac{1}{2\pi i}\int_\gamma \lambda[e^{-\lambda(t-\tau+\varepsilon)} - e^{-\lambda(t-\tau'+\varepsilon)}](\lambda I - A)^{-1}\,d\lambda] \\ &\leq \frac{1}{2\pi}\cdot 2\int_0^\infty r\cdot \mathrm{re}^{-r\cos\theta(t+\varepsilon-\tau^*)}\cdot|\tau-\tau'|\cdot\frac{C}{r\cos\theta}\,dr,\end{aligned} \tag{3.18}$$

where $\tau^* \in S_h$. Since $t - \tau^* \geq (t-\tau)/3$, from (3.18) we obtain

$$|K_\varepsilon(t,\tau) - K_\varepsilon(t,\tau')| \leq C_1 h\cdot\int_0^\infty \mathrm{re}^{-1/3r\cos\theta(t-\tau)}\,dr = \frac{C_2 h}{(t-\tau)^2}. \tag{3.19}$$

Estimating the integral (3.17) with the help of (3.19), we find that

$$\begin{aligned}I &\leq C_2 h\int_{t_0+2h}^{T}\frac{dt}{(t-\tau)^2} \\ &\leq C_2 h\int_{t_0+2h}^{\infty}\frac{dt}{(t-\tau)^2} \\ &= \frac{C_2 h}{t_0+2h-\tau} \leq C_2.\end{aligned} \tag{3.20}$$

Here we have used the fact that for $\tau \in S_h$ the denominator $t_0 + 2h - \tau \geq h$.

Theorem 3.1 now follows from the extrapolation Theorem 1.5.

THEOREM 3.2. *Suppose the operator A satisfies condition* (2.2) *and its spectrum does not contain points of the imaginary axis. Then the estimates of solutions of the Cauchy problem and of the periodic problem are equivalent. More precisely, if for a given $p > 1$ inequality* (3.4) *holds at least for one time segment* $[0, T]$, *then for a T-periodic solution for any T the estimate* (3.5) *holds and C does not depend on T. If for at least one period*

T_0 the estimate of the periodic solution (3.5) *holds, then for any $T > 0$* (3.4) *is satisifed.*

PROOF. It may be assumed that $a = 0$. We further remark that from the estimate (3.4) for some one value $T = T_0$ we obtain it for any $T > 0$. Indeed, for $T < T_0$ it suffices to extend $f(t)$ by setting it equal to 0 for $t > T_0$. If $T > T_0$, then we decompose $[0, T]$ into a sum of segments of length no more than T_0, and we write (3.4) for each of them. To estimate terms arising as a result of inhomogeneity of the initial condition we use an inequality of the form

$$|||u(t)|||_p \leq \left\| \frac{du}{dt} \right\|_{L_p(T_1 T_2)} + \|Au\|_{L_p(T_1,T_2)} \qquad (0 < T_1 \leq \tau \leq T_2). \tag{3.21}$$

By summing, we get (3.4) for any T.

Solving the Cauchy problem for the segment $[0, T]$, we further agree to assume that the right side $f(t)$ has been continued T-periodically to the entire t-axis.

Comparing the solution $u(t)$ of the Cauchy problem (3.9) and the periodic solution $u_0(t)$ of (2.51), we obtain

$$\begin{aligned} u_0(t) &= u(t) + v_1(t) - v_2(t); \\ v_1(t) &= \int_{-\infty}^{0} e^{-(t-\tau)A_+} P_+ f(\tau)\, d\tau; \\ v_2(t) &= \int_{0}^{\infty} e^{-(t-\tau)A_-} P_- f(\tau)\, d\tau. \end{aligned} \tag{3.22}$$

We shall show that

$$\|Av_k\|_{L_p(0,T)} \leq C\|f\|_{L_p(0,T)} \qquad (k = 1, 2). \tag{3.23}$$

Assuming that $\operatorname{Re}\sigma_+(A) > \sigma_1 > 0$ and applying Lemma 2.1, we find

$$|Av_1(t)|_X \leq \int_{-\infty}^{0} \frac{Ce^{-\sigma_1(t-\tau)}}{t-\tau} |f(\tau)|_X\, d\tau. \tag{3.24}$$

Using the periodicity of $f(\tau)$, we represent (3.24) in the form

$$|Av_1(t)|_X \leq \sum_{k=1}^{\infty} \int_0^T \frac{e^{-\sigma_1(t+kT+s)}}{t+kt+s} |f(-s)|_X\, ds. \tag{3.25}$$

From (3.25), dropping kt in the denominator and summing, we obtain

$$|Av_1(t)| \leq \int_0^T \frac{|f(-s)|_X}{t+s}\, ds. \tag{3.26}$$

The estimate (3.23) for $k = 1$ follows immediately from (3.26) by Theorem 1.4 on the boundedness in L_p of the Hilbert integral operator. The estimate

(3.23) for $k = 2$ can be derived in a similar manner (actually, this estimate is even simpler, since $\sigma_-(A)$ is a bounded set).

Theorem 3.2 follows directly from (3.22) and (3.23).(3)

2. Some equations in Hilbert space. In the general case the theorems of the preceding subsection do not make it possible to arrive at definitive results. It would be very tempting to use the representation of a periodic solution of (3.1) in the form of a Fourier series:

$$Au_0(t) = \sum_{k=-\infty}^{+\infty} A(i\omega kI + A)^{-1} f_k e^{ik\omega t};$$

$$f = \sum_{k=-\infty}^{+\infty} f_k e^{ik\omega t}; \qquad \omega = \frac{2\pi}{T}. \tag{3.27}$$

The multipliers $\Lambda_k = A_k(i\omega kI + A)^{-1}$ satisfy the "Mikhlin condition"

$$|\Lambda_k| \le C; \qquad \left|\frac{d\Lambda_k}{dk}\right| \le \frac{C}{1+|k|}. \tag{3.28}$$

However, the Marcinkiewicz theorem has been proved only in the case of Hilbert space and is unknown even for $X = L_p(\Omega)$. This annoying circumstance explains the presence in Chapter I of §7, in which the required estimate is derived for the Navier-Stokes equation.

In this subsection coerciveness inequalities are proved for one important class of equations in a Hilbert space H. In particular, this class contains many parabolic equations (see §4) and the Navier-Stokes equations (§5) if solutions are considered whose leading derivatives are square-summable.

LEMMA 3.1. *Suppose an operator $A\colon X \to X$ generates a semigroup which is analytic in some sector (i.e., satisfies condition (2.2)). Suppose a closed operator $R\colon X \to X$ is such that $D_R \supset D_A$ and*

$$\|R(\lambda I - A)^{-1}\| \to 0, \tag{3.29}$$

when $|\lambda| \to \infty$, where λ runs through points of the sector $\Sigma_{\sigma_0,\varphi}$. Then the operator $A + R$ also generates a semigroup analytic in a sector (i.e., a condition of the form (2.2) holds for it, possibly with different σ_0 and C, but with the same φ).

PROOF. We have

$$(\lambda I - A - R)^{-1} = R_\lambda(I - RR_\lambda)^{-1}; \qquad R_\lambda = (\lambda I - A)^{-1}. \tag{3.30}$$

(3)That C is independent of T is proved in the Appendix to §5.

By (3.29), for some $r > 0$ and any $\lambda \in \Sigma_{\sigma_0,\varphi}$ such that $|\lambda| \geq r$ we have

$$\|RR_\lambda\| < \frac{1}{2}. \tag{3.31}$$

For such λ from (3.30) we obtain

$$\|(\lambda I - A - R)^{-1}\| \leq 2\|R_\lambda\| \leq \frac{2C}{|\lambda - \sigma_0| + 1}. \tag{3.32}$$

We now shift the sector $\Sigma_{\sigma_0,\varphi}$ to the left so that the disk $|\lambda| \leq r$ remains outside it. For this it suffices to place its vertex at the point $\sigma_1 = \sigma_0 - r/\sin\varphi$. Thus, if $\lambda \in \Sigma_{\sigma_0,\varphi}$ condition (3.32) holds.

We shall further show that the operator $A + R$ is closed. Let u_n $(n = 1, 2, \dots)$ be a sequence of elements in D_A converging in the norm of X, and suppose that $(A + R)u_n$ is also a convergent sequence: $u_n \to u$ and $(A+R)u_n = f_n \to f$. It is necessary to show that $u \in D_A$ and $(A+R)u = f$. We set $(\lambda I - A)u_n = v_n$, where λ satisfies condition (3.31). The sequence $\{v_n\}$ then converges, since

$$v_n = (I - RR_\lambda)^{-1}(\lambda u_n - f_n).$$

Since the operator $\lambda I - A$ is closed, it now follows that $u \in D_A$ and

$$(\lambda I - A)u = v = (I - RR_\lambda)^{-1}(\lambda u - f).$$

We then obtain

$$(A + R)u = \lambda u - (I - RR_\lambda)v = f,$$

as required. The lemma is proved.

We remark that condition (3.29) is sufficient in order that the operator R have fractional degree α $(0 < \alpha < 1)$ relative to A.

THEOREM 3.3. *Suppose A is a linear operator acting in a Hilbert space H and representable in the form $A_0 + R$, where A_0 is self-adjoint and positive definite and R satisfies the condition*

$$\|R(\lambda I \to A_0)^{-1}\| \to 0, \quad \textit{where } \lambda \to -\infty. \tag{3.33}$$

Then for the solution $u(t)$ of the Cauchy problem

$$\frac{du}{dt} + (A_0 + R)u = f(t); u(0) = 0 \tag{3.34}$$

there is the estimate $(0 \leq T \leq +\infty; p > 1)$

$$\left(\int_0^T e^{p\sigma_0 t}\left(\|A_0 u\|_H^p + \left\|\frac{du}{dt}\right\|_H^p dt\right)\right)^{1/p} \leq \frac{Cp^2}{p-1}\left(\int_0^T e^{p\sigma_0 t}\|f(t)\|_H^p\, dt\right)^{1/p}, \tag{3.35}$$

We consider the mixed problem for a parabolic equation

$$\frac{\partial u}{\partial t} + Au = f(x,t); \tag{4.1}$$

$$B_j u|_S = 0; \tag{4.2}$$

$$u|_{t=0} = a(x). \tag{4.3}$$

Here $u = u(x,t)$ is an unknown function and A is the differential operator

$$Au = \sum_{|\alpha|\leq 2m} a_\alpha(x) D^\alpha u, \tag{4.4}$$

where $\alpha = (\alpha_1, \dots, \alpha_n)$ is a vector index with natural number components and $|\alpha| = \alpha_1 + \cdots + \alpha_n$. D^α denotes the operator

$$D^\alpha u = \frac{\partial^{\alpha_1}}{\partial x_1^{\alpha_1}}, \frac{\partial^{\alpha_2}}{\partial x_2^{\alpha_2}}, \dots, \frac{\partial^{\alpha_n}}{\partial x_n^{\alpha_n}} u. \tag{4.5}$$

B_j is a differential boundary operator of order $m_j < 2m$.

$$B_j u = \sum_{|\alpha|\leq m_j} b_{j,\alpha}(x) D^\alpha u; \qquad x \in S. \tag{4.6}$$

We shall assume that the operator A with the boundary conditions (4.2) is regular elliptic in the sense of Agmon, Douglis, and Nirenberg [1]. This means, first of all, that the form of degree $2m$

$$A'(x,\xi) \equiv \sum_{|\alpha|=2m} a_\alpha(x) \xi_1^{\alpha_1} \xi_2^{\alpha_2} \cdots \xi_n^{\alpha_n} > 0 \tag{4.7}$$

for all $x \in \overline{D}$ and any real $\xi_1, \dots, \xi_n$; $\sum_1^n \xi_k^2 > 0$. Second, it is necessary that the polynomials $B_j(x, \xi + s\nu)$ in S ($x \in S$; ξ is a vector tangent to S at the point x) be linearly independent modulo the polynomial $\prod_1^m (s - s_k(\xi))$, where the s_k are the roots of the polynomial $A'(x, \xi + s\nu)$ with positive imaginary part.

The next assertion, in particular, follows from the results of [2].

THEOREM 4.1. *Suppose the differential operator A with boundary conditions* (4.2) *is regular elliptic, $S \in C^{2m}$, and the coefficients of the leading derivatives in* (4.1) *are continuous, while the remaining are bounded and measurable. Suppose the coefficients of the operators B_j are continuously differentiable $2m - m_j$ times. Then in $L_p(\Omega)$ the operator A has discrete spectrum, its resolvent set contains the sector $\Sigma'_{\sigma_0,\varphi} = \{\sigma : \pi \geq |\arg(\sigma - \sigma_0)| \geq \varphi\}$ (σ_0 is some real number; $0 < \varphi < \pi/2$), and*

$$\|(\sigma I - A)^{-1}\|_{L_p \to W_p^{(2m)}(\Omega)} \leq C; \tag{4.8}$$

$$\|(\sigma I - A)^{-1}\|_{L_p \to L_p} \leq \frac{C_p}{|\sigma| + 1}$$

for any $\sigma \in \Sigma'_{\sigma_0,\varphi}$; $p > 1$, and C_p does not depend on σ.

A regular elliptic operator thus satisfies condition (2.3) of §2.

LEMMA 4.1. *Suppose A is a regular elliptic operator of order $2m$ in a bounded domain Ω; consider it as an operator in $L_p(\Omega)$ $(p > 1)$. Then the operator of generalized differentiation $L\colon L_p(\Omega) \to L_r(\Omega)$, $Lu = D_x^k u$ $(0 \leq k \leq 2m-1;\ r \geq p)$, has degree*

$$\alpha = \frac{1}{2m}\left(\frac{n}{p} - \frac{n}{r} + k\right) \tag{4.9}$$

relative to the operator A; r and k are taken so that $\alpha < 1$.

PROOF. From the results of [60], [19], and [26]–[28] we obtain

$$\|Lu\|_{L_r(\Omega)} \leq C\|u\|^{\alpha}_{W_p^{(2m)}(\Omega)} \cdot \|u\|^{1-\alpha}_{p(\Omega)}, \tag{4.10}$$

where C does not depend on $u \in W_p^{(2m)}$. Setting $u = (\sigma I - A)^{-1} f$ and using (4.8), for any $\sigma \in \Sigma_{\sigma_0,\varphi}$ we find that

$$\|L(\sigma I - A)^{-1}\|_{L_p \to L_p} \leq \frac{C}{|\sigma|^{1-\alpha}}, \tag{4.11}$$

as required.

By applying this lemma and using the results of §2, we arrive at the following theorem.

THEOREM 4.2. *Suppose $\alpha \in L_p(\Omega)$ and the function f is continuous in x and t $(x \in \overline{\Omega}, t \geq 0)$. Then problem (4.1)–(4.3) has a unique generalized solution $u(x,t)$, which satisfies*

$$\|u(\cdot,t)\|_{L_p(\Omega)} \leq Ce^{-\sigma_0 t}\left\{\|a\|_{L_p(\Omega)} + \left[\int_0^t (e^{\sigma_0\tau}\|f(\cdot,\tau)\|_{p_1(\Omega)})^r\, d\tau\right]^{1/r}\right\}$$

$$\left(p > 1;\quad r > 1 \text{ any number};\quad \frac{1}{p_1} = \frac{2m}{n}\frac{1}{r'} + \frac{1}{p}\right); \tag{4.12}$$

$$\begin{aligned}\left[\int_0^t (e^{\sigma_0 r}\|D_x^k u(\cdot,\tau)\|_{L_p(\Omega)})^q\, d\tau\right]^{1/q} \\ \leq C\left\{e^{-\sigma_0 t}\|a\|_{L_{p_1}(\Omega)} + \left[\int_0^t e^{\sigma_0 l\tau}\|f(\cdot,\tau)\|^l_{p_2(\Omega)}\, d\tau\right]^{1/l}\right\},\end{aligned} \tag{4.13}$$

$$\frac{1}{q} = \frac{1}{2m}\left(\frac{n}{p_1} - \frac{n}{p} + k\right);$$

$$\frac{1}{p_2} = \frac{2m}{n}\left(1 - \frac{1}{l} + \frac{1}{q}\right) - \frac{1}{n}\left(k - \frac{n}{p}\right);$$

$$p \geq p_1 > 1; \qquad q > p_1 > 1.$$

We note a special case of (4.13): $q = p$, $l = p_2$; and

$$\frac{1}{p} = \frac{1}{2m+n}\left(\frac{n}{p_1} + k\right); \quad \frac{1}{l} = \frac{1}{p} + \frac{2m-k}{2m+n}; \qquad \left(k \leq \frac{2m}{p_1}\right).$$

If $q \leq p_1$, then an estimate of the form (4.13) holds with the replacement $p_1 \to p_1 + \varepsilon$ ($\varepsilon > 0$ is arbitrary).

It is of interest to obtain an estimate of the type (4.12) for $D_x u$. We write the solution of problem (4.1)–(4.3) in the form $u = u_1 + u_2$, where u_1 is the solution in the case $f = 0$ and u_2 is the solution in the case $a = 0$. For u_2 we deduce from Lemma 2.7 that

$$\|D_x^k u_2(\cdot, t)\|_{L_p(\Omega)} \leq Ce^{-\sigma_0 t}\left[\int_0^t e^{r\sigma_0\tau}\|f(\cdot,\tau)\|^r_{L_{p_1}(\Omega)}\,d\tau\right]^{1/r}; \quad (4.14)$$

$$\left(\frac{n}{p} = \frac{n}{p_1} + k - \frac{2m}{r'}; \quad 0 < \frac{n}{p_1} + k - \frac{2m}{r'} < n; \quad r < p; \quad p_1 < p\right).$$

If $r \geq p$, then a weaker estimate of the type (4.14) holds with the replacement $r \to r + \varepsilon$.

We shall briefly describe the method of estimating the term u_1. We denote C_A^{2m} the subspace in C^{2m} consisting of functions satisfying the boundary conditions (4.2), and by $W_{p,A}^{(k)}$ the closure of C_A^{2k} in the norm of $W_p^{(k)}$. Among other things, it follows from the imbedding theorems that functions in $W_{p,A}^{(k)}$ satisfy in the classical sense those of the boundary conditions (4.2) for which $m_j < k - n/p$, and in mean those for which $k - n/p \leq m_j \leq n - 1 + k - n/p$.

It turns out that the operator A generates an analytic semigroup not only in L_p but also in $W_{p_0,A}^{(k)}$. This assertion is deduced from the following lemma, which we present here without proof.

LEMMA 4.2. *Consider the boundary value problem*

$$\sigma u - Au = Af; \quad (4.15)$$

with the boundary conditions (4.2). *Suppose $S \in C^k$. Then for any $k \geq 0$ the estimate* (*which is uniform with respect to $\sigma \in \Sigma_{\sigma_0,\theta}$*)

$$\|u\|_{W_p^{(k)}(\Omega)} \leq C\|f\|_{W_p^{(k)}(\Omega)}. \quad (4.16)$$

holds.

This lemma (and even a more general one with A replaced by another operator A_1 of order $2m$) can be proved just as its analogue for $\sigma = 0$ is proved in [1].

We assume that the point $\sigma = 0$ does not belong to the spectrum of A (otherwise we would go over to the operator $A - \mu I$). We represent any function $a \in W_p^{(k)}$ as a sum $a = a_1 + a_2$, where a_1 is the solution of boundary value problem (4.2) for the equation $Aa_1 = Aa$. According to the lemma, the projection operator P_1 defined by the equality $P_1 a = a_1$ is bounded in $W_p^{(k)}$, and it maps $W_p^{(k)}$ into $W_{p,A}^{(k)}$. In exactly the same way the projection $P_2 = I - P_1$ is bounded in $W_p^{(k)}$ and maps $W_p^{(k)}$ into the closure of the set of solutions of the homogeneous differential equation $Au = 0$.

We now define an operator A_0 in $W_{p,A}^{(k)}$ with domain $W_{p,A}^{(2m)}$ by setting $A_0 = P_1 A$. We shall show that A_0 is the generator of an analytic semigroup in $W_p^{(k)}$. With a view to obtaining an estimate of the resolvent of A_0 we consider the equation

$$(\sigma I - A_0)u = f, \tag{4.17}$$

where $\sigma \in \Sigma_{\sigma_0,\theta}$ and $f \in W_{p,A}^{(k)}$; since $W_{p,A}^{(2m)}$ is dense in $W_{p,A}^{(k)}$, it may be assumed that $f \in W_{p,A}^{(2m)}$. Equation (4.17) is equivalent to

$$(\sigma I - A)u = f, \tag{4.18}$$

since $P_2 f = P_2 u = P_2 A u = 0$. Setting $Au = v$, we now see that v is the solution of boundary value problem (4.2) for the equation

$$(\sigma I - A)v = Af. \tag{4.19}$$

Applying results of [1] to estimate u in terms of v, and the lemma to estimate v in terms of f, we obtain

$$\|u\|_{W_{p,A}^{(k+2m)}} \leq C_1 \|v\|_{W_p^{(k)}} \leq C\|f\|_{W_p^{(k)}}. \tag{4.20}$$

After this, from (4.19) we obtain

$$\|u\|_{W_{p,A}^{(k)}} \leq \frac{C}{|\sigma|} \|f\|_{W_{p,A}^{(k)}} \tag{4.21}$$

and the assertion is proved.

As a result of the above considerations we arrive at the following theorem.

THEOREM 4.3. *Suppose $S \in C^k$. Then the solution of problem* (4.1)–(4.3) *satisfies*

$$\|D_x^k u(\cdot,t)\|_{L^p(\Omega)} \le C\Big\{ e^{-\sigma_0 t}\|a\|_{W_{p,A}^{(k)}} + \Big[\int_0^t e^{r\sigma_0\tau}\|f(\cdot,\tau)\|_{L_{p_1}(\Omega)}^r\,d\tau\Big]^{1/r}\Big\}; \tag{4.22}$$

$$\Big(\frac{n}{p} = \frac{n}{p_1} + k - \frac{2m}{r'};\quad 0 < \frac{n}{p'} + k - \frac{2m}{r'} < n;\quad 0 \le k < 2m\Big).$$

If $f = 0$, it is possible to take $k = 2m$.

We now proceed to estimates of the leading derivatives. We define the space $V_{p,r}^{(l)}$ consisting of functions $a(x)$ ($x \in \Omega$) such that there exists a function $v(x,t)$ defined in the cylinder $Q = \Omega \times [0,\infty]$ satisfying the boundary conditions (4.2) for $t > 0$, where $v(x,0) = a(x)$ and v has the finite norm

$$\|v\|_{H_{p,r}^{(l)}}^r = \sum_{k=0}^{l}\int_0^\infty e^{r\sigma_0 t}\left\|\frac{\partial^k v}{\partial t^k}\right\|_{L_p(\Omega)}^r dt + \sum_{k=0}^{2lm}\int_0^\infty e^{r\sigma_0 t}\|D_x^k v\|_{L_p(\Omega)}^r\,dt. \tag{4.23}$$

The norm in $V_{p,r}^{(l)}$ is defined by

$$\|a\|_{V_{p,r}^{(l)}} = \inf_{v(x,0)=a}\|v\|_{H_{p,r}^{(l)}}. \tag{4.24}$$

THEOREM 4.4. *Suppose $S \in C^{2m}$. Then*

$$\|u\|_{H_{p,r}^{(1)}} \le C(\|a\|_{V_{p,r}^{(0)}} + \|f\|_{H_{p,r}^{(0)}}) \qquad (p, r > 1). \tag{4.25}$$

PROOF. Since for $r = p$ the estimate (4.25) is known, Theorem 4.4 follows immediately from Theorem 3.1 and 4.1.

Estimates of the derivatives of any order can also be derived from Theorem 4.4. Indeed, we set $u_k = \partial^k u/\partial t^k$ and $f_k = \partial^k f/\partial t^k$. From (4.1)–(4.3) we then obtain for u_k the conditions

$$\frac{\partial u_k}{\partial t} + Au_k = f_k; \tag{4.26}$$

$$B_j u_k|_S = 0; \tag{4.27}$$

$$u_k|_{t=0} = \sum_{r=0}^{k-1}(-1)^r A^r f_{k-r-1} + (-1)^k A^k a \equiv a_k. \tag{4.28}$$

From Theorem 4.4 we then obtain

$$\|u_k\|_{H_{p,r}^{(1)}} \le C(\|a_k\|_{V_{p,r}^{(0)}} + \|f\|_{H_{p,r}^{(0)}}). \tag{4.29}$$

We suppose that $\operatorname{Re}\sigma(A) < 0$ (we can go over to this case by the change $u_k = e^{\mu t}u'_k$). From (4.26) for any $t > 0$ we then obtain

$$\|u_k\|_{W_p^{(2m)}(\Omega)} \leq C(\|f\|_{L_p(\Omega)} + \|u_{k+1}\|_{L_p(\Omega)}). \tag{4.30}$$

Combining (4.29) and (4.30), we arrive at the following result.

THEOREM 4.5. *Suppose $S \in C^{2lm}$ and $a_1 \in V_{p,r}^{(1)}, \dots, a_k \in V_{p,r}^{(k)}$. Then the solution of problem* (4.1)–(4.3) *satisfies*

$$\|u\|_{H_{p,r}^{(l)}} \leq C\left(\sum_{k=0}^{l} \|a_k\|_{V_{p,r}^{(k)}} + \|f\|_{H_{p,r}^{(k-1)}}\right). \tag{4.31}$$

We remark that the condition that the function a_k belongs to the space $V_{p,r}^{(l)}$ implies, together with smoothness, also consistency conditions–algebraic relations between the derivatives of the functions f and a for $t = 0$ and $x \in S$.

If it is required only that $a \in L_p$, one can obtain estimates of the derivatives of the solution $u(x,t)$ in the cylinder $\Omega'_\eta = \Omega \times [\eta, \infty]$, $\eta > 0$. Indeed, suppose $\psi(t)$ is an infinitely differentiable function such that $\psi(t) = 1$ for $t \geq \eta$ and $\psi(t) = 0$ for $t \leq \eta/2$. The function $\tilde{u} = \psi u$ then satisfies

$$\frac{\partial \tilde{u}}{\partial t} - A\tilde{u} = \psi f + \psi'(t)u. \tag{4.32}$$

Applying the preceding theorem to it, we get the following assertion.

THEOREM 4.6. *Suppose $S \in C^{2lm}$ and $a \in L_{p_1}(\Omega)$ $(p_1 > 1)$. Then* $(p > 1; k, r \geq 1)$

$$\int_\eta^\infty e^{r\sigma_0 t}\left(\left\|\frac{\partial^k u}{\partial t^k}\right\|_{L_p(\Omega)}^r + \|D_x^{2km}u\|_{L_p(\Omega)}^r\right) dt \leq C(\|a\|_{L_{p_1}}^r + \|f\|_{H_{p,r}^{(k-1)}}^r) \tag{4.33}$$

Finally, we formulate a result pertaining to bounded solutions and, in particular, periodic and almost periodic solutions.

THEOREM 4.7. *Suppose the spectrum of the operator A does not contain points of the imaginary axis, and suppose $f(x,t)$ is a bounded measurable function for $-\infty < t < \infty$ with values in $L_p(\Omega)$ $(p > 1)$:*

$$\sup_{-\infty<t<\infty} \|f(\cdot,t)\|_{L_p(\Omega)} < \infty \tag{4.34}$$

Then problem (4.1), (4.2) *has a unique generalized solution $u(x,t)$ which is bounded in t $(-\infty < t < \infty)$ in the sense that*

$$\sup_t \|D_x^k u(\cdot,t)\|_{L_r(\Omega)} \leq C \sup_t \|f(\cdot,t)\|_{L_p(\Omega)}; \tag{4.35}$$

$$\left(0 < \frac{1}{2m}\left(\frac{n}{p} - \frac{s}{r} + k\right) < 1\right).$$

If f is T-periodic or almost periodic in t, then $u(x,t)$ has the same property.

This theorem follows immediately from Theorem 2.1 and Lemma 4.1; other estimates of bounded solutions can be obtained from Theorems 2.2 and 3.2. In particular, the next result follows from Theorem 3.2 and 4.4

THEOREM 4.8. *Suppose $f(x, t)$ is periodic in t, while the spectrum of the operator A does not contain the numbers $2k\pi i/T$ $(k = 0, \mp 1, \dots)$. Then a T-periodic solution of problem* (4.1), (4.2) *satisfies*

$$\int_0^T \left(\left\| \frac{\partial u}{\partial t} \right\|^r_{L_p(\Omega)} + \|D_x^{2m} u\|^r_{L_p(\Omega)} \right) dt \le C \int_0^T \|f\|^r_{L_p(\Omega)}\, dt, \quad (4.36)$$

$$(p, r > 1).$$

Local existence theorems for quasilinear parabolic equations can be derived from the results of this section.

§5. The linearized Navier-Stokes equations

In this section we begin the study of the Navier-Stokes equations in the case of an incompressible fluid filling some domain Ω with boundary S on which the velocity vector $\overline{v} = \overline{v}(x, t)$ (x is a point of the domain Ω and t is time) assumes a prescribed value. Mathematically, the situation reduces to the problem

$$\frac{\partial v}{\partial t} + (\overline{v}, \nabla)\overline{v} - \nu \Delta \overline{v} = -\nabla P + \overline{F}; \quad (5.1)$$

$$\operatorname{div} \overline{v} = 0; \quad (5.2)$$

$$\overline{v}|_S = \overline{\alpha}; \quad (5.3)$$

$$\overline{v}|_{t=0} = \overline{v}_0. \quad (5.4)$$

Here P is the pressure in the fluid, $\nu > 0$ is the kinematic coefficient of viscosity which is assumed to be constant, the density of the fluid is taken equal to 1, $\overline{F} = \overline{F}(x, t)$ is a given vector of external mass forces, and $\overline{\alpha} = \overline{\alpha}(x, t)$ and $\overline{v}_0(x)$ are given vectors.

We suppose that $\overline{\alpha}$ and $\overline{F}$ do not depend on the time t, and that problem (5.1)–(5.4) has a steady-state solution $(\overline{a}(x), P_0(x))$. Our immediate goal is to study the Navier-Stokes equations linearized in a neighborhood of this solution:

$$\frac{\partial \overline{u}}{\partial t} - \nu \Delta \overline{u} + (\overline{a}, \nabla)\overline{u} + (\overline{u}, \nabla)\overline{a} = -\nabla q + \overline{F}; \quad (5.5)$$

$$\operatorname{div} \overline{u} = 0; \quad (5.6)$$

$$\overline{u}|_S = 0; \quad (5.7)$$

$$\overline{u}|_{t=0} = \overline{u}_0(x). \quad (5.8)$$

We shall assume that Ω is a bounded domain of three-dimensional space whose boundary S consists of a finite number of closed surfaces of class C^2. The results carry over to the two-dimensional problem in an entirely automatic manner (even with some simplifications). The assumption of boundedness is also not essential in many cases, but giving it up would greatly complicate the exposition.

We begin with some definitions. We introduce the set M of smooth solenoidal vectors (satisfying (5.6) in Ω) which have a zero normal component on S.

We denote by S_p ($p \geq 1$) the Banach space obtained by closing the set M in the norm of L_p:

$$\|\overline{u}\|_{S_p} = \left[\int_\Omega |\overline{u}(x)|^p \, dx\right]^{1/p} = \|\overline{u}\|_{L_p}. \tag{5.9}$$

Further, suppose that G_p is the Banach space obtained by closing the set of gradients of smooth (single-valued) functions in the norm of L_p:

$$\|\nabla\varphi\|_{G_p} = \left[\int_\Omega |\nabla\varphi|^p \, dx\right]^{1/p} = \|\nabla\varphi\|_{L_p}. \tag{5.10}$$

The next lemma shows that S_p and G_p for $p > 1$ are not only linear manifolds but subspaces of L_p, and L_p decomposes into the direct sum

$$L_p = S_p \oplus G_p. \tag{5.11}$$

We remark that for $p = 1$ (as well as for $p = \infty$) this is not the case.

Lemma 5.1. *Let Π be the orthogonal projection in L_2 onto S_2. Then Π acts boundedly in L_p; more precisely, it admits a continuation from the set L (dense in L_p) of smooth vectors with normal component vanishing on S to a bounded operator in L_p. Moreover,*

$$\|\Pi\|_{L_p \to S_p} \leq C\frac{p^2}{p-1}, \tag{5.12}$$

where the constant C depends on S but not on p.

Proof. Let $\overline{b}$ be a smooth vector vanishing on S. We define a function φ as the solution of the Neumann problem

$$\Delta\varphi = \operatorname{div}\overline{b} \equiv \sum_{k=1}^{3} \frac{\partial b_k}{\partial x_k}; \left.\frac{\partial\varphi}{\partial n}\right|_S = 0. \tag{5.13}$$

According to Theorem 1.7,

$$\|\nabla\varphi\|_{L_p} \leq C\frac{p^2}{p-1}\|\overline{b}\|_{L_p}. \tag{5.14}$$

We introduce the vector $\overline{a}$ by setting

$$\overline{b} = \overline{a} + \operatorname{grad} \varphi. \tag{5.15}$$

It is obvious that $\nabla\varphi \in G_p$ and $\overline{a} \in S_p$. We define the operator Π by setting $\overline{a} = \Pi\overline{b}$. Clearly $\Pi^2 = \Pi$, i.e., Π is a projection; the fact that it realizes an orthogonal projection in L_2 follows immediately from the following relation, which can be verified by integration by parts:

$$\begin{aligned}(\operatorname{grad} \varphi, \overline{a})_{L_2} &= \int_\Omega \operatorname{grad} \varphi \cdot \overline{a}^* \, dx = \int_\Omega \operatorname{div}(\varphi\overline{a}^*)\, dx \\ &= \int_S \varphi a_n^* \, ds = 0 \end{aligned} \tag{5.16}$$

for any smooth function φ and $a \in M$. The estimate (5.12) follows immediately from (5.14) and (5.15). The lemma is proved.

The operator Π, introduced in hydrodynamics by S. G. Kreĭn [39] and E. Hopf [24], has a clear physical meaning. If we consider a fluid as a system of material points subject to the constraint (5.2) (the condition of incompressibility), then it can be treated as the orthogonal projection onto the "surface" cut out by the constraint. Under such a projection, of course, the reaction of an (ideal) constraint (5.2) vanishes. There can be no doubt that the operator Π will subsequently appear not only in theoretical investigations but also in the solution of concrete problems, and not only in intermediate arguments but also in the final answers. For example, it has already appeared (without being named) in criteria for stability of steady flow, in formulas for the curvature tensor of the group of diffeomorphisms preserving the volume, etc. (see Arnol′d [6]).

LEMMA 5.2. *The set M_0 of smooth solenoidal vectors vanishing on S is dense in S_p $(p \geq 1)$.*

PROOF. It suffices to establish that any vector $\overline{a} \in M$ can be approximated to arbitrary accuracy in the norm of L_p by vectors in M_0. We shall prove that a vector $\overline{a} \in M$ can be represented in the form

$$\overline{a} = \operatorname{curl} \overline{B}; \overline{B}/s = 0. \tag{5.17}$$

Indeed, it is known that a smooth solenoidal vector $\overline{a}$ in the domain Ω with zero flux across the boundary S can be represented in the form

$$\overline{a} = \operatorname{curl} \overline{B}_0; \qquad B_{0\mathbf{n}}/_S = 0; \tag{5.18}$$

$$\oint_{\gamma_k} \overline{B}_0 \cdot d\overline{x} = 0 \qquad (k = 1, \dots, r), \tag{5.19}$$

where $\gamma_1, \dots, \gamma_r$ is a complete collection of independent closed contours in Ω. The tangential vector field $\overline{B}_0 = \overline{B}_0(x)$ on S is a potential field, because for any closed contour γ lying on S and bounding some region $\Sigma \subset S$ we have

$$\oint_\gamma \overline{B}_0 \cdot \overline{dx} = \int_\Sigma \operatorname{curl} \overline{B_0} \cdot \overline{n}\, ds = \int_\Sigma a_n\, ds = 0, \tag{5.20}$$

since $a_n|_S = 0$. Moreover, by (5.19) and (5.20) the vector $\overline{B}_0$ has zero circulation about any closed contour $\gamma \subset S$, and hence its potential is well defined. Thus, on the boundary S

$$\overline{B}_0 = \operatorname{grad} \varphi_0(s); \qquad s \in S, \tag{5.21}$$

where φ_0 is a (single-valued!) function defined on S. We extend it with preservation of smoothness to the interior of Ω. This extension can be carried out, for example, in the following manner. Let $\rho(x)$ denote the distance from a point x to S, and let Ω_h be an h-neighborhood of the boundary S: $\Omega_h = \{x \in \Omega \colon \rho(x) < h\}$. Suppose that h is so small that the normals to S do not intersect in Ω_h (it suffices to take h less than the minimal radius of curvature of the boundary S). In Ω_h it is then possible to introduce "curvilinear coordinates" (s, ρ), where $s = s(x)$ is a point of S whose distance from x is minimal and $\rho = \rho(x)$ is that distance. The desired extension can be defined by

$$\varphi(x) = \begin{cases} \dfrac{[\rho(x) - h]^2}{h^3}[2\rho(x) + h]\varphi_0(s(x)) & (x \in \Omega_h); \\ 0 & (x \in \Omega = \Omega_h). \end{cases} \tag{5.22}$$

It is clear that $\varphi \in C^1$, $\varphi|_S = \varphi_0$, and $\partial\varphi/\partial n|_S = 0$. In order that (5.17) be satisfied it now suffices to set

$$\overline{B} = \overline{B}_0 - \nabla\varphi. \tag{5.23}$$

From (5.17) we obtain

$$|B(x)| \le C\rho(x). \tag{5.24}$$

Suppose now that $\psi(\tau)$ is an infinitely differentiable function decreasing monotonically on $[0, \infty]$ and such that

$$\psi(\tau) = \begin{cases} 1 & \left(0 \le \tau \le \frac{1}{2}\right), \\ 0 & (\tau \ge 1). \end{cases} \tag{5.25}$$

We define the function $\eta_h(x)$ $(x \in \overline{\Omega})$ by

$$\eta_h(x) = 1 - \psi\left(\frac{\rho(x)}{2h}\right). \tag{5.26}$$

The function $\eta_n \in C^1$ is nonnegative and does not exceed 1. We introduce the vector $\overline{a}_h$ by

$$\overline{a}_h(x) = \operatorname{curl}(\eta_h \overline{B}) = \eta_h \overline{a} + \nabla \eta_h \times \overline{B}. \tag{5.27}$$

The vector $\overline{a}_h$ lacks only smoothness to belong to the set M_0 (it is only continuous). We remedy this situation by mollification. Let

$$\begin{aligned} \overline{a}_{h\delta}(x) &= \int_\Omega K_\delta(x-y)\overline{a}_h(y)\,dy \\ &= \int_{R^3} K_\delta(x-y)\overline{a}_h(y)\,dy, \end{aligned} \tag{5.28}$$

where K_δ is a mollification kernel of radius $\delta < h$; for example, it is possible to take

$$K_\delta(x) = \frac{1}{K_\delta^0}\psi\left(\frac{|x|}{\delta}\right); \qquad K_\delta^0 = \delta^3 \int_{R^3} \psi(|x|)\,dx. \tag{5.29}$$

Outside the domain Ω we set the vector $\overline{a}_h$ equal to 0. The vector $\overline{a}_{h\delta}$ is infinitely differentiable, equal to 0 in $\Omega_{h-\delta}$, and is solenoidal, since by (5.27) and (5.28)

$$\overline{a}_{h\delta} = \operatorname{curl} \int_{R^3} K_\delta(x-y)\eta_h(y)\overline{B}(y)\,dy. \tag{5.30}$$

Thus, $\overline{a}_{h\delta} \in M_0$. Further, we have

$$\|\overline{a} - \overline{a}_{h\delta}\|_{L_p} \leq \|\overline{a} - \overline{a}_{h\delta}\|_{L_p} + \|\overline{a}_h - \overline{a}_{h\delta}\|_{L_p}. \tag{5.31}$$

Estimating the first term on the right side of (5.31) and considering (5.27), we obtain

$$\|\overline{a} - \overline{a}_h\|_{L_p(\Omega)} \leq \|\overline{a}\|_{L_p(\Omega_{2h})} + \|\nabla \eta_h \times \overline{B}\|_{L_p(\Omega_{2h})}. \tag{5.32}$$

Further, using (5.24) and the estimate $|\nabla \eta_h| < C/h$, where C does not depend on h, we find that

$$\|\nabla \eta_h \times \overline{B}\|_{L_p(\Omega_{2h})} \leq C h^{1/p}. \tag{5.33}$$

It is now clear that the right side in (5.32) vanishes as $h \to 0$. Let $\varepsilon > 0$ be any number. We fix h so that

$$\|\overline{a} - \overline{a}_h\|_{L_p} < \varepsilon/2. \tag{5.34}$$

From the theory of mollification kernels it follows that for sufficiently small δ

$$\|\overline{a}_h - \overline{a}_{h\delta}\|_{L_p} < \varepsilon/2. \tag{5.35}$$

From (5.31), (5.34) and (5.35) we obtain

$$\|\overline{a} - \overline{a}_{h\delta}\|_{L_p} < \varepsilon. \tag{5.36}$$

Lemma 5.2 is thus proved.

In S_p $(p > 1)$ we now define the operators A_0, A, and R by setting, for any solenoidal vector $\overline{u} \in W_p^{(2)}(\Omega)$ vanishing on S,

$$\begin{aligned} A_0\overline{u} &= -\Pi\Delta\overline{u}; \\ A\overline{u} &= \nu A_0\overline{u} + R\overline{u}; \\ R\overline{u} &= \Pi[(\overline{a}, \nabla)\overline{u} + (\overline{u}, \nabla)\overline{a}]. \end{aligned} \tag{5.37}$$

We denote the domains of A, A_0, and R in S_p by $D_p(A), D_p(A_0)$, and $D_p(R)$.

Problem (5.5)–(5.7) can now be treated as a Cauchy problem for the ordinary differential equation in the Banach space S_p

$$\frac{d\overline{u}}{dt} + A\overline{u} = \overline{f}; \qquad \overline{u}(0) = \overline{u}_0. \tag{5.38}$$

Here $\overline{f} = \Pi\overline{F}$. The pressure q is found from the relation

$$\nabla q = -\Pi' \left\{ \frac{\partial\overline{u}}{\partial t} - v\Delta\overline{u} + (\overline{a}, \nabla)\overline{u} + (\overline{u}, \nabla)\overline{a} - \overline{F} \right\}, \tag{5.39}$$

where $\Pi' = I - \Pi$ is the orthogonal projection in L_2 onto G_2. The function q is necessarily single-valued by the definition of the spaces G_p; $\partial u/\partial t$ in (5.39) can obviously be dropped.

LEMMA 5.3. *The operator A_0 in S_p $(p > 1)$ is closed, while in S_2 it is self-adjoint and positive definite.*

PROOF. We choose any sequence $\overline{u}_n \in D_p(A_0)$ convergent in S_p; let $\overline{u}$ be its limit. We suppose that the sequence $A_0\overline{u}_n$ converges in S_p: $A_0\overline{u}_n \to \overline{g}$. It is necessary to show that $\overline{u} \in D_p(A_0)$ and $A_0\overline{u} = \overline{g}$. Now from Theorem 6.1 it follows that $\overline{u}_n \to \overline{u}$ in $W_p^{(2)}$. Therefore, closedness of the operator A_0 follows from the closedness of operators of generalized differentiation.

The symmetry of the operator A_0 in S_2 follows from the identity

$$\begin{aligned} (A_0\overline{u}, \overline{v})_{S_2} &= -\int_\Omega \nu\Pi\Delta\overline{u} \cdot \overline{v}^*\, dx \\ &= \nu \int_\Omega \Delta\overline{u} \cdot \overline{v}^*\, dx \\ &= \nu \int_\Omega \operatorname{curl}\overline{u} \cdot \operatorname{curl}\overline{v}^*\, dx \\ &= \int_\Omega \sum_{k=1}^3 \frac{\partial\overline{u}}{\partial x_k}\frac{\partial\overline{v}^*}{\partial x_k}\, dx \\ &= (\overline{u}, A_0, \overline{v})_{S_2}, \end{aligned} \tag{5.40}$$

which for any $\overline{u}, \overline{v} \in D_2(A)$ can be verified by integration by parts. Further, from (5.40) and the Friedrichs inequality for any $\overline{u} \in D_A$ we deduce that

$$\begin{aligned}(A_0\overline{u}, \overline{u})_{S_2} &= \int_\Omega \sum_{k=1}^{3} \left|\frac{\partial \overline{u}}{\partial x_k}\right|^2 dx \\ &\geq \gamma^2 \int_\Omega |\overline{u}|^2\, dx \\ &= \gamma^2 \|\overline{u}\|^2_{S_2},\end{aligned} \tag{5.41}$$

where $\gamma > 0$ depends only on the domain Ω but not on $\overline{u}$.

Thus, A_0 is a positive-definite operator in S_2. It is self-adjoint, since its range is the entire space S_2. (See [52], §5.) The lemma is proved.

LEMMA 5.4. *Any bounded linear functional l defined on S_p $(p > 1)$ can be represented in the form*

$$l\overline{u} = \int_\Omega \overline{u} \cdot \overline{v}^*\, dx, \tag{5.42}$$

where $\overline{v} \in S_{p'}$ $(1/p + 1/p' = 1)$ and is uniquely determined by the functional l. Here

$$\|l\| \leq \|\overline{v}\|_{S_{p'}} \leq C\frac{p^2}{p-1}\|l\|. \tag{5.43}$$

PROOF. By the Hahn-Banach theorem the functional l can be extended with preservation of norm to all of $L_p = L_p(\Omega, m_3, R^3 + iR^3)$. By the Riesz theorem it can be represented in the form

$$l\overline{u} = \int_\Omega \overline{u} \cdot \overline{v}_0^*\, dx, \tag{5.44}$$

where $\overline{v}_0 \in L_{p'}$ and $\|\overline{v}_0\|_{L_{p'}} = \|l\|$. Setting now $\overline{v} = \Pi\overline{v}_0$ and applying Lemma 5.1, we obtain (5.42) and (5.43). If there existed a representation of l in the form (5.42) with another vector $\overline{v}_1 \in S_{p'}$ in place of $\overline{v}$, then for $\overline{v} = \overline{v} - \overline{v}_1$ for all $\overline{u} \in S_p$ we would have

$$\int_\Omega \overline{u} \cdot \overline{v}'^*\, dx = 0. \tag{5.45}$$

However, assuming $\overline{u} = \Pi(|\overline{v}'|^{p-2}\overline{v}')$ in (5.45), we find that $\|\overline{v}'\|_{S_{p'}} = 0$, and so $\overline{v}' = 0$. The lemma is proved.

It follows from this lemma that the mapping $l \to \overline{v}$ realized by (5.42) is an isomorphism of the dual space S_p^* and the space $S_{p'}$; in this sense it is possible to assume that $S_p^* = S_{p'}$.

Applying Lemma 5.4, repeating the computation (5.40), and again using the estimate in L_p of the leading derivatives of solutions of the Navier-Stokes system, we conclude that the adjoint operator A_0^* of the operator A_0

in S_p is the same operator A_0 but considered in $S_{p'}$. As shown below, the operator A is also closed, R admits closure (it suffices to set, by definition, $R\overline{u} = R^{**}\overline{u}$), while the adjoint operators R^* and A^* are

$$R^*\overline{u} = -\Pi\left\{a_r\left(\frac{\partial v_s}{\partial x_r} + \frac{\partial v_r}{\partial x_s}\right)\overline{i}_s\right\}; \qquad A^* = \nu A_0 + R^*. \tag{5.46}$$

The domain of A^* is the same as for A_0 in $S_{p'}$, i.e., $D_p(A^*) = D_p(A)$, $p > 1$. The operator R^* is given by (5.46) only for smooth vectors. The general definition consists in the following. A vector $\overline{u} \in S_{p'}$ belongs to the domain $D_{p'}(R^*)$ of the operator R^* if (and only if) it can be assigned a vector $\overline{g} \in S_{p'}$ such that

$$\int_\Omega R\overline{\Phi}\cdot\overline{u}^*\,dx = \int_\Omega \overline{\Phi}\cdot\overline{g}^*\,dx; \qquad \overline{\Phi} \in D_p(A). \tag{5.47}$$

Here, by definition, $R^*\overline{u} = g$.

We remark that the operators R and A in S_2 are nonsymmetric whatever the (solenoidal) vector $\overline{a} \neq 0$ [99].

We now study the spectrum and resolvent of the operator A. We begin with A_0. We consider the equation

$$(\sigma I - A_0)\overline{u} = \overline{f}, \tag{5.48}$$

where σ is a complex parameter and $\overline{f} \in S_p$. We introduce the energy space H_1 of the operator A_0. H_1 is the Hilbert space obtained by closing $D_2(A_0)$ in the metric

$$\begin{aligned}(\overline{u},\overline{v})_{H_1} &= (A_0 u, v)_H \\ &= (A_0^{1/2}u, A_0^{1/2}v)_H \\ &= \int_\Omega \sum_{k=1}^3 \frac{\partial\overline{u}}{\partial x_k}\frac{\partial\overline{v}^*}{\partial x_k}\,dx.\end{aligned} \tag{5.49}$$

The space H_1 consists of vectors of class $W_2^{(1)}$ which are solenoidal and vanish on S; according to the Sobolev imbedding theorem, H_1 is continuously imbedded in L_6, while in L_p ($p < 6$) it is compactly imbedded, and

$$\|\overline{v}\|_{L_p} \leq C_p\|\overline{v}\|_{H_1}, \tag{5.50}$$

where C_p depends only on the domain Ω, while C_6 may be assumed to be an absolute constant.

A generalized solution of (5.48) is a vector $\overline{u} \in H_1$ such that

$$\sigma(\overline{u},\overline{\Phi}) - (\overline{u},\overline{\Phi})_{H_1} = (\overline{f},\overline{\Phi}) \tag{5.51}$$

for all $\overline{\Phi} \in H_1$. In order that the right side be meaningful, according to (5.50), it suffices to assume that $\overline{f} \in L_{6/5}$. It may also be assumed that $\overline{f}$ is a generalized function of the form $\Pi(\sum_1^3 \partial f_k/\partial x_k)$; $\overline{f}_k \in L_2$. Here, by definition, for $\overline{\Phi} \in H_1$ we set

$$(\overline{f}, \overline{\Phi}) = -\int_\Omega \sum_{k=1}^3 \overline{f}_k \cdot \frac{\partial \overline{\Phi}^*}{\partial x_k}\, dx. \tag{5.52}$$

We define an operator $G: L_{6/5} \to H_1$ by requiring that

$$-(G\overline{f}, \overline{\Phi})_{H_1} = (\overline{f}, \overline{\Phi}); \qquad \overline{\Phi} \in H_1. \tag{5.53}$$

Since the right side of (5.53) is a continuous linear functional on H_1, the existence and unique determination of $G\overline{f}$ follow from the theorem of F. Riesz on the general form of a linear functional in Hilbert space. It is further clear that the definition of generalized solution in (5.51) is equivalent to the following operator equation in H_1:

$$\overline{u} = \sigma G \overline{u} + \overline{u}_0; \qquad \overline{u}_0 = -G\overline{f}. \tag{5.54}$$

As is known [45], strict positivity of the operator G follows from positive definiteness of A_0. The compactness of the imbedding of H_1 in L_p ($p < 6$) noted above implies the compactness of G in H_1 and in H. Therefore, the spectrum of G in H_1 (and in H) consists of a sequence of positive eigenvalues $1/\sigma_{01}, 1/\sigma_{02}, \dots$. Here $0 < \sigma_{01} \le \sigma_{02} \le \cdots$, and $\sigma_{0k} \to +\infty$. The corresponding eigenvectors $\overline{\psi}_{01}, \overline{\psi}_{02}, \dots$ form a complete system in H and in H_1.

From Theorem 6.1 (see §6 below) it follows that for $p > 6/5$ the generalized solution $\overline{u}$ of (5.54) lies in $D_p(A_0)$. Thus, $G = A_0^{-1}$, and the resolvent $R_\sigma(A_0) = (\sigma I - A_0)^{-1}$ exists for all $\sigma \ne \sigma_{01}, \sigma_{02}, \dots$ The same is true for $1 < p \le 6/5$: for the proof it suffices to go over the adjoint operator and observe that

$$\|R_\sigma(A_0)\|_{S_p \to S_p} = \|R_\sigma(A_0)\|_{S_{p'} - S_{p'}}.$$

When p runs over the ray $p \ge 6/5$ the dual index $p' = p/(p-1)$ runs over the segment $(1, 6)$.

We further remark that the smoothness of the vectors $\{\overline{\psi}_{0k}\}$ is determined only by the smoothness of the boundary S. Indeed, if, for example, it is first supposed that $\overline{\psi}_{0k} \in H$, then from the equation

$$\overline{\psi}_{0k} = \sigma_{0k} G \overline{\psi}_{0k} \tag{5.55}$$

by results of [83] (see also Theorem 6.1) and the imbedding theorem it follows that $\overline{\psi}_{0k} \in D_2(A_0) \subset W_2^{(2)} \subset C(\overline{\Omega})$. Now, again applying Theorem

6.1, we find that $\overline{\psi}_{0k} \in W_p^{(2)}$ for any $p > 1$. If $S \in C^l$, then $\overline{\psi}_{0k} \in W_p^{(l)}$ for any $p > 1$.

We now say a few words regarding the character of completeness of the system $\{\overline{\psi}_{0k}\}$. Suppose $\overline{u} \in D_2(A_0)$. Then the vector $A_0\overline{u} \in H$, and it can be approximated by a linear combination $\overline{\psi} = \sum_1^N \alpha_k \overline{\psi}_{0k}$ such that

$$\left\| A_0\overline{u} - \sum_{k=1}^{N} \alpha_k \overline{\psi}_k \right\|_H < \varepsilon, \tag{5.56}$$

where $\varepsilon > 0$ is any number. But then by Theorem 6.1

$$\left\| \overline{u} - \sum_{k=2}^{N} \frac{\alpha_k}{\sigma_{0k}} \overline{\psi}_{0k} \right\|_{W_2^{(2)}} \le C \left\| A_0 u - \sum_{k=1}^{N} \alpha_k \overline{\psi}_{0k} \right\|_H \le C\varepsilon. \tag{5.57}$$

Thus, the system $\{\overline{\psi}_{0k}\}$ is complete in $D_2(A_0)$ in the metric of $W_2^{(2)}$. By the imbedding theorem it is then also complete in $D_2(A_0)$ in the metric of S_p for any $p > 1$. Since $D_2(A_0)$ is dense in S_p, the system $\{\overline{\psi}_{0k}\}$ is complete in S_p. Replacing H by S_p in the foregoing argument, we conclude that it is also dense in $D_p(A_0)$ $(p > 1)$ relative to the metric of $W_p^{(2)}$.

We formulate these conclusions as a lemma.

LEMMA 5.5. *The operator A_0 in S_p $(p > 1)$ has pure point spectrum consisting of an infinite sequence of positive eigenvalues $0 < \sigma_{01} \le \sigma_{02} \le \cdots$; $\sigma_{0k} \to +\infty$. The corresponding eigenvectors $\{\overline{\psi}_{0k}\}$ form a complete system in S_p and also in $D_p(A_0)$ in the metric of $W_p^{(2)}$. Moreover, if $S \in C^l$, then $\overline{\psi}_{0k} \in W_p^{(l)}$ for any $p > 1$.*

REMARK. Using the familiar procedure of Carleman (see [71]), one can study the asymptotics of the eigenvalues σ_{0k} for large k.(4) It turns out to be almost the same as for the Laplace operator (in the latter case the coefficient 6 occurs in place of 3):

$$\sigma_{0k} \sim \left(\frac{3\pi^2 k}{m_3 \Omega} \right)^{2/3}; \qquad k \to \infty. \tag{5.58}$$

In the two-dimensional case the answer is the same as for the Laplace operator:

$$\sigma_{0k} \sim \frac{4\pi k}{m_2 \Omega}; \qquad k \to \infty. \tag{5.59}$$

Proceeding to the study of the operator A, we consider in S_p $(p > 1)$ the equation

$$(\sigma I - A)\overline{u} = \overline{f}. \tag{5.60}$$

---

(4)Metivier (1978) obtained this asymptotics by a variational method.

A generalized solution of this equation is a vector $\overline{u} \in H_1$ satisfying

$$\sigma(\overline{u},\overline{\Phi}) - \nu(\overline{u},\overline{\Phi})_{H_1} - (R\overline{u},\overline{\Phi}) = (\overline{f},\overline{\Phi}); \qquad \overline{\Phi} \in H_1 \tag{5.61}$$

By (5.53) an equivalent definition is the following: the vector $\overline{u} \in H_1$ is a solution of the equation

$$\left(\frac{\sigma}{\nu}G - I - \frac{1}{\nu}GR\right)\overline{u} = \overline{u}_0; \qquad \overline{u}_0 = \frac{1}{\nu}G\overline{f}. \tag{5.62}$$

Assuming first that $\overline{f} \in S_{6/5}$, we conclude that $\overline{u}_0 \in H_1$. Further, the operator GR acts compactly in H_1. Indeed, it is obvious from (5.37) that R acts continuously from H_1 to S_2, while G takes S_2 into $W_2^{(2)}$ which is compactly imbedded in $W_2^{(1)}$.

We shall prove the last assertion in another manner which is suitable for any bounded domain (no conditions are imposed on the smoothness of the boundary).

We take an arbitrary sequence converging weakly in H_1: $\overline{u}_n \to \overline{u}$. By the imbedding theorem it converges in the norm of S_p $(p < 6)$. From the definition (5.53) of the operator G we have

$$-(GR\overline{u}_n,\overline{\Phi})_{H_1} = (R\overline{u}_n,\overline{\Phi}); \qquad \overline{\Phi} \in H_1. \tag{5.63}$$

From (5.63) we deduce that for any $\overline{\Phi} \in H_1$

$$-(GR\overline{u}_n - GR\overline{u}_m,\overline{\Phi})_{H_1} = (R\overline{u}_n - R\overline{u}_m,\overline{\Phi}). \tag{5.64}$$

Setting $\overline{\Phi} = GR\overline{u}_n - GR\overline{u}_m$ in (5.64) and applying the Cauchy-Schwarz-Bunyakovskiĭ inequality and (5.50), we obtain

$$\begin{aligned}\|GR\overline{u}_n - GR\overline{u}_m\|_{H_1}^2 &= (\overline{u}_n - \overline{u}_m, R^*GR(\overline{u}_n - \overline{u}_m)) \\ &\leq \|R^*G\|_{H\to H} \cdot \|\overline{u}_n - \overline{u}_m\|_H \cdot \|R(\overline{u}_n - \overline{u}_m)\|_H \to 0,\end{aligned} \tag{5.65}$$

since the first and third factors are bounded, while the second tends to 0 as $m,n \to \infty$. It has thus been proved that the operator GR is strongly continuous and hence also compact in H_1.

We further remark that the operator GR admits extension to a compact operator in H: It suffices to set, by definition, $GRu = (R^*G)^*\overline{u}$ for any $\overline{u} \in H$.

LEMMA 5.6. *The operator A is closed and has pure point spectrum which consists of an infinite sequence of eigenvalues $\sigma_1, \sigma_2, \ldots$. Here* $\operatorname{Re}\sigma_k \to +\infty$ *as $k \to \infty$ (hence, there exist no more than a finite number of eigenvalues* *with negative real part). The spectrum of A is contained in the region*([5])

$$\operatorname{Re}\sigma \geq -\gamma + \nu\sigma_{01}; \qquad |\operatorname{Im}\sigma| \leq \alpha\sqrt{\frac{\operatorname{Re}\sigma + \gamma}{\nu}} + \beta; \tag{5.66}$$
$$\alpha = \max_{x\in\Omega} |\overline{a}(x)|; \qquad \beta = \max_{x\in\Omega} |\operatorname{curl}\overline{a}(x)|;$$
$$\gamma = \max_{x\in\Omega} \sqrt{\sum_{i,k=1}^{3} \left(\frac{\partial a_i}{\partial x_k}\right)^2}.$$

The sequence of eigenvectors and associated vectors of the operator A forms a complete system in H and in H_1, and also in S_p and $D_p(A)$ ($p > 1$) in the sense of the metric of $W_p^{(2)}$.

PROOF. Suppose $\overline{u}_n \in D_p(A)$, and $\overline{u}_n \to \overline{u}$ and $A\overline{u}_n \to \overline{g}$ in the norm of S_p. We shall show then that $\overline{u} \in D_p(A)$ and $A\overline{u} = \overline{g}$. We introduce the notation $A\overline{u}_n = \overline{g}_n$ and obtain

$$\left(I - \frac{1}{\nu}GR\right)\overline{u}_n = \frac{1}{\nu}G\overline{g}_n. \tag{5.67}$$

The operator GR can be extended by continuity of all of S_p by setting $GR\overline{u} = (R^*G)^*\overline{u}$ for any $\overline{u} \in S_p$. Now from Lemma 5.5 and the imbedding theorem it follows that R^*G acts continuously from S_p to $W_p^{(1)} \subset L_p$ for any $p > 1$. Hence, GR acts continuously in S_p. It follows from Theorem 6.2 that GR acts boundedly from S_p to $W_p^{(1)}$. Passing to the limit in (5.67), we obtain

$$\nu\overline{u} = GR\overline{u} + G\overline{g}. \tag{5.68}$$

The right side in (5.68) belongs to $W_p^{(1)}$, and hence $\overline{u} \in W_p^{(1)}$. Then $R\overline{u} + \overline{g} \in S_p$ and by (5.68)

$$\overline{u} = (1/\nu)G(R\overline{u} + \overline{g}) \in D_p(A_0).$$

It is now possible to apply the operator A_0 to (5.68) and find that $A\overline{u} = \overline{g}$. Thus, A is closed in S_p ($p > 1$). Closedness of A in S_p can be proved without appealing to Theorem 6.2. Indeed, as shown above, the operator GR acts boundedly from S_p ($p \geq 2$) to H_1. Thus, if $\overline{u} \in S_p$ is a solution of (5.68), then $\overline{u} \in H_1$. Further, by Lemma 5.5, GR acts from H_1 to $W_2^{(2)} \subset W_6^{(1)}$. Hence, $\overline{u} \in W_6^{(2)} \subset W_p^{(1)}$. Finally, GR acts from $W_p^{(1)}$ to $D_p(A)$. From (5.68) it now follows that $A\overline{u} = \overline{g}$, and so A is closed for $p \geq 2$; in the case $p < 2$ it suffices to observe that $A = A^{**}$.

([5])We point out that $\operatorname{Re}\sigma \geq -\gamma$ uniformly with respect to ν. In the case of an ideal fluid ($\nu = 0, a_n|_S = 0$) the spectrum lies in the strip $|\operatorname{Re}\sigma| \leq \gamma$ but, of course, it must no longer be a pure point spectrum.

The arguments just made show that the equation in S_p

$$(\sigma I - A)\overline{u} = \overline{f} \tag{5.69}$$

is equivalent to

$$\left(\frac{\sigma}{\nu}G - I - \frac{1}{\nu}GR\right)\overline{u} = \frac{1}{\nu}G\overline{f}. \tag{5.69'}$$

Hence, the resolvent sets and spectra of the operator A and of the pencil (5.69) coincide.

The operator $(GR)^2$ acts boundedly from H to $W_2^{(2)} \subset C$, and hence [16] it is a Hilbert-Schmidt operator in H. From results of Keldysh [30] (see also [20]) we therefore conclude that the system of eigenvectors and associated vectors of the pencil of operators $(\sigma G - I - GR)$ is complete in H, and the spectrum of the pencil is discrete; it consists of eigenvalues $\sigma_1, \sigma_2, \dots$, and $\operatorname{Re}\sigma_k \to +\infty$. This result was formulated by S. G. Kreĭn [39] (see also [90]).

It can be proved exactly as in Lemma 5.5 that the spectrum and root vectors of the pencil $\sigma G - I - GR$ in S_p do not depend on p, while the latter possess the smoothness of $W_p^{(2)}$ for any $p \geq 1$; if $S \in C^l$ they belong to $W_p^{(l)}$ for all $p \geq 1$.

We shall prove that the system of eigenvectors and associated vectors is complete in $D_p(A)$ in the sense of the metric of $W_p^{(2)}$. It is known that the system of eigenvectors $\overline{\psi}_k = \overline{\psi}_k^{(0)}$ can be chosen so that the associated vectors $\overline{\psi}_k^{(1)}, \dots, \overline{\psi}_k^{(m_n)}$ are determined by the equations

$$(\sigma_k I - A)\psi_k^{(m)} = \psi_k^{(m-1)} \qquad (m = 1, 2, \dots, m_k). \tag{5.70}$$

If $\overline{u} \in D_2(A)$, then we take $\sigma_0 \in \sigma(A)$ and find a linear combination

$$\chi = \sum_{k=1}^{N} \sum_{m=0}^{m_k} c_{km}\overline{\psi}_k^{(m)}$$

which approximates $(\sigma_0 I - A)\overline{u}$ well in the norm of H. Then $(\sigma_0 I - A)^{-1}\chi$ approximates $\overline{u}$ well in the norm of $W_2^{(2)}$. Now $(\sigma_0 I - A)^{-1}\chi$ is obviously a finite linear combination of the vectors $\psi_k^{(m)}$. Completeness in $D_2(A)$ has thus been proved. Since $D_2(A)$ is dense in S_p and the norm of S_p is subordinate to the norm of $W_2^{(2)}$ for any $p \geq 1$, completeness in any S_p $(p \geq 1)$ also holds. Completeness in $D_p(A)$ relative to the metric of $W_p^{(2)}$ is now deduced from completeness in S_p exactly as it was done above for $p = 2$.

We now prove (5.66). Taking the inner product of the equation

$$\left(\frac{\sigma}{\nu}G - I - \frac{1}{\nu}GR\right)\overline{u} = 0 \tag{5.71}$$

with $\overline{u}$ in H_1 and noting (5.37) and (5.53), we find that

$$\nu\|\overline{u}\|_{H_1}^2 - \sigma\|\overline{u}\|_H^2 = -\int_\Omega [(\overline{a},\nabla)\overline{u} + (\overline{u},\nabla)\overline{a}]\overline{u}^*\,dx$$
$$= \int_\Omega [\overline{a}\times\operatorname{curl}\overline{u} + \overline{u}\times\operatorname{curl}\overline{a}]\overline{u}^*\,dx. \tag{5.72}$$

Separating the real and imaginary parts, we obtain

$$\nu\|\overline{u}\|_{H_1}^2 - \operatorname{Re}\sigma\|\overline{u}\|_H^2 = -\operatorname{Re}\int_\Omega (\overline{u}\cdot\nabla)\overline{a}\cdot\overline{u}^*\,dx; \tag{5.73}$$
$$\operatorname{Im}\sigma\|\overline{u}\|_H^2 = \operatorname{Im}\int_\Omega [\overline{a}\times\operatorname{curl}\overline{u}\cdot\overline{u}^* + \overline{u}\times\operatorname{curl}\overline{a}\cdot\overline{u}^*]\,dx.$$

From (5.73) we derive the estimates

$$\nu\|\overline{u}\|_{H_1}^2 \le (\operatorname{Re}\sigma + \gamma)\|\overline{u}\|_H^2; \tag{5.74}$$
$$|\operatorname{Im}\sigma|\cdot\|\overline{u}\|_H^2 \le \alpha\|\overline{u}\|_{H_1}\cdot\|\overline{u}\|_H + \beta\|\overline{u}\|_H^2. \tag{5.75}$$

From the minimal property of the first eigenvalue of the operator A_0 it follows that

$$\|\overline{u}\|_H^2 \le \sigma_{01}^{-1}\|\overline{u}\|_{H_1}^2. \tag{5.76}$$

If $\overline{u}\ne 0$, then from (5.74) and (5.76) we obtain $\operatorname{Re}\sigma+\gamma\ge\nu\sigma_{01}$. Further, from (5.74) we obtain

$$\|\overline{u}\|_{H_1} \le \sqrt{\frac{\operatorname{Re}\sigma+\gamma}{\nu}}\|\overline{u}\|_H. \tag{5.77}$$

The second inequality of (5.66) now follows from (5.75) and (5.77).

Lemma 5.6 is proved.

REMARK. From the results of [30] it is easy to derive the asymptotics of the eigenvalues of the operator A. Namely, it turns out that $\sigma_k/\nu\sigma_{0k}\to 1$ as $k\to\infty$. If we note the remark to Lemma 5.5, then we obtain in the three-dimensional case

$$\sigma_k \sim \nu\left(\frac{3\pi^2 k}{m_3\Omega}\right)^{2/3}; \qquad k\to\infty,$$

and in the two-dimensional case

$$\sigma_k \sim \nu\frac{4\pi k}{m_2\Omega}; \qquad k\to\infty.$$

We now choose any real number $\sigma_0 < -\gamma+\nu\sigma_{01}$. By (5.66) the resolvent set of the operator A contains the sector

$$\Sigma_{\sigma_0,\theta} = \{\sigma\colon \theta\le|\arg(\sigma-\sigma_0)|\le\pi\},$$

where θ belongs to the interval $(0,\pi/2)$ and is sufficiently close to $\pi/2$.

LEMMA 5.6. *The following estimates hold uniformly with respect to* $\sigma \in \Sigma_{\sigma_0,\theta}$ *and* $\overline{f} \in H$:

$$\|(\sigma I - A)^{-1}\overline{f}\|_H \leq \frac{C}{|\sigma - \sigma_0|}\|\overline{f}\|_H; \tag{5.78}$$

$$\|(\sigma I - A)^{-1}\overline{f}\|_{H_1} \leq \frac{C}{\sqrt{|\sigma - \sigma_0}}\|\overline{f}\|_H. \tag{5.79}$$

PROOF. Suppose $\overline{u}$ is a solution in S_2 of the equation

$$(\sigma I - A)\overline{u} = \overline{f}. \tag{5.80}$$

Below we shall use only the fact that $\overline{u}$ is a generalized solution of (5.80), i.e., $\overline{u} \in H_1$ and the following equation is satisfied:

$$(\sigma G - I - GR)\overline{u} = G\overline{f}. \tag{5.81}$$

Taking the inner product of (5.81) with $\overline{u}$ in H_1, separating real and imaginary parts, and making the same estimates as in the derivation of (5.74) and (5.75), we obtain

$$\nu\|\overline{u}\|^2_{H_1} \leq (\operatorname{Re}\sigma + \gamma)\|\overline{u}\|^2_H + \|\overline{f}\|_H \cdot \|\overline{u}\|_H; \tag{5.82}$$

$$|\operatorname{Im}\sigma| \cdot \|\overline{u}\|^2_H \leq \alpha\|\overline{u}\|_{H_1}\|\overline{u}\|_H + \beta\|\overline{u}\|^2_H + \|\overline{f}\|_H \cdot \|\overline{u}\|_H. \tag{5.83}$$

We make the estimate separately in two cases:

a) $\operatorname{Re}\sigma \geq \sigma_0$; $|\operatorname{Im}\sigma| \geq |\operatorname{Re}(\sigma - \sigma_0)\tan\theta$;

b) $\operatorname{Re}\sigma \leq \sigma_0$.

In case a), estimating the right side of (5.83) with application of the elementary inequality $ab < (\varepsilon^2a^2 + b^2/\varepsilon^2)/2$ $(a, b, \varepsilon > 0)$ and noting (5.82), we find that

$$\left[|\operatorname{Im}\sigma| - \frac{\alpha\varepsilon^2}{2\nu}(\operatorname{Re}\sigma + \gamma) - \frac{\alpha}{2\varepsilon^2} - \beta\right]\|u\|_H \leq \left(1 + \frac{\alpha\varepsilon^2}{2\nu}\right)\|\overline{f}\|_H \tag{5.84}$$

We now use inequality a) and the following consequence of it:

$$|\operatorname{Im}\sigma| \geq |\sigma - \sigma_0|\sin\theta. \tag{5.85}$$

From (5.84) we then obtain

$$\|\overline{u}\|_H \leq \frac{1 + \frac{\alpha\varepsilon^2}{2\nu}}{\left(1 - \frac{\alpha\varepsilon^2}{2\nu}\cot\theta\right)\sin\theta \cdot |\sigma - \sigma_0| - \frac{\alpha\varepsilon^2}{2\nu}(\sigma_0 + \gamma) - \frac{\alpha}{2\varepsilon^2} - \beta}\|\overline{f}\|_H. \tag{5.86}$$

Here ε is assumed so small that $1 - (\alpha\varepsilon^2/2\nu)\cot\theta > 0$, while $|\sigma - \sigma_0|$ is so large that the denominator in (5.86) is positive.

In case b) inequality (5.84) is preserved; from it, by applying b), for sufficiently large $|\operatorname{Im}\sigma|$ we conclude that

$$\|\overline{u}\|_H \leq \frac{1+\frac{\alpha\varepsilon^2}{2\nu}}{|\operatorname{Im}\sigma| - \frac{\alpha\varepsilon^2}{2\nu}(\sigma_0+\gamma) - \frac{\alpha}{2\varepsilon^2} - \beta}\|\overline{f}\|_H. \tag{5.87}$$

From (5.82), by applying (5.76), we obtain

$$\begin{aligned} \|\overline{u}\|_H &\leq \frac{1}{\nu\sigma_{01} - \operatorname{Re}\sigma - \gamma}\|\overline{f}\|_H \\ &\leq \frac{1}{|\operatorname{Re}(\sigma-\sigma_0)|}\|\overline{f}\|_H. \end{aligned} \tag{5.88}$$

It clearly suffices to prove (5.78) for large $|\sigma - \sigma_0|$. In case a) it follows directly from (5.86). In case b) it follows from (5.87) for any vertical strip $\sigma_1 \leq \operatorname{Re}\sigma \leq \sigma_0$ and from (5.88) for any horizontal strip $|\operatorname{Im}\sigma| \leq \rho$. It remains to show that it holds when σ varies inside the quadrants $\operatorname{Re}(\sigma - \tau) < 0, |\operatorname{Im}\sigma| \geq \rho$.

Here it may be assumed that $|\tau|$ and ρ are sufficiently large fixed numbers, and $\tau < \sigma_0$. Now in this case from (5.87) and (5.88) we obtain

$$\|\overline{u}\|_H \leq \frac{C}{\max\{|\operatorname{Re}(\sigma-\sigma_1)|, |\operatorname{Im}(\sigma-\sigma_1)|\}}\|\overline{f}\|_H; \qquad \sigma_1 = \tau + i\rho, \tag{5.89}$$

which is equivalent to (5.78), since

$$\begin{aligned} \max\{|\operatorname{Re}(\sigma-\sigma_1)|, |\operatorname{Im}(\sigma-\sigma_1)|\} &\geq \frac{1}{\sqrt{2}}|\sigma-\sigma_1| \\ &\geq |\sigma - \sigma_0|. \end{aligned}$$

This concludes the derivation of (5.78). Inequality (5.79) follows immediately from (5.78) and (5.82). The lemma is proved.

Lemma 5.7. *Let σ_0 be some real number such that $\sigma_0 < -\gamma + \nu\sigma_{01}$. Then the resolvent of the operator A in S_p $(p > 1)$ satisfies*

$$\|(\sigma I - A)^{-1}\|_{S_p \to S_p} \leq \frac{C}{|\sigma - \sigma_0|}; \tag{5.90}$$

$$\|(\sigma I - A)^{-1}\|_{S_p \to W_p^{(2)}} \leq C \tag{5.91}$$

uniformly with respect to $\sigma \in \Sigma_{\sigma_0,\theta} = \{\sigma \colon \theta \leq |\arg(\sigma - \sigma_0)| \leq \pi\}$ where $\theta \in (0, \pi/2)$ and is sufficiently close to $\pi/2$.

Proof. It follows from Lemma 5.6 that the sector $\Sigma_{\sigma_0,\theta}$ lies entirely in the resolvent set of A. Since (5.90) obviously follows from (5.91), we limit ourselves to a derivation of the latter. Applying the multiplicative inequality, we obtain for any $\overline{u} \in D_p(A)$

$$\|Ru\|_{S_p} \leq C\|u\|_{S_p}^{1/2} \cdot \|u\|_{W_p^{(2)}}^{1/2}. \tag{5.92}$$

Setting $\overline{u} = (\sigma I - \nu A_0)^{-1}\overline{f}, \overline{f} \in S_p$, in (5.92) and using Theorem 6.1, we arrive at the following estimate which is uniform with respect to $\sigma \in \Sigma_{\sigma_0,\theta}$:

$$\|RR_{0\sigma}\|_{S_p \to S_p} \leq \frac{C}{\sqrt{|\sigma|}}; \qquad R_{0\sigma} = (\sigma I - \nu A_0)^{-1}. \tag{5.93}$$

It suffices to prove (5.91) for large $|\sigma|$; for example, we shall assume that $|\sigma|$ is so large that

$$\|RR_{0\sigma}\|_{S_p \to S_p} \leq \tfrac{1}{2}. \tag{5.94}$$

It then remains to write the resolvent of A in the form

$$(\sigma I - A)^{-1} = R_{0\sigma}(I - RR_{0\sigma})^{-1} \tag{5.95}$$

and note that by Theorem 6.1 and condition (5.94)

$$\begin{aligned}\|(\sigma I - A)^{-1}\|_{S_p \to W_p^{(2)}} &\leq \|R_{0\sigma}\|_{S_p \to W_p^{(2)}} \cdot \|(I - RR_{0\sigma})^{-1}\|_{S_p \to S_p} \\ &\leq 2\|R_{0\sigma}\|_{S_p \to W_p^{(2)}} \leq C.\end{aligned}$$

The lemma is proved.

This lemma incorporates the linearized Navier-Stokes equations into the general theory developed above. It now remains to formulate theorems pertaining to equation (5.38) to which problem (5.5)–(5.8) will be reduced.

THEOREM 5.1. *The operator A in S_p for any $p > 1$ generates an analytic semigroup. For the solution of the Cauchy problem* (5.38) *the estimates*

$$\|\overline{u}(t)\|_{S_p} \leq Ce^{-\sigma_0 t}\left\{\|\overline{u}_0\|_{S_p} + \left[\int_0^t (e^{\sigma_0\tau}\|\overline{f}(\tau)\|_{S_{p_1}(\Omega)})^{r_1}\, d\tau\right]^{1/r_1}\right\}$$

$$\frac{1}{p_1} = \frac{2}{n}\frac{1}{r_1'} + \frac{1}{p}; \qquad r_1 \leq p;$$

$$\begin{aligned}&\left[\int_0^t (e^{\sigma_0\tau}\|D_x^k\overline{u}(\tau)\|_{L_p})^q\, d\tau\right]^{-1/q} \\ &\qquad \leq C\left\{\|\overline{u}_0\|_{S_{p_2}} + \left[\int_0^t e^{\sigma_0 r_2\tau}\|\overline{f}(\tau)\|_{S_{p_3}}^{r_2}\, d\tau\right]^{1/r_2}\right\};\end{aligned}$$

$$\frac{1}{q} = \frac{1}{2}\left(\frac{n}{p_2} - \frac{n}{p} + k\right);$$

$$\frac{1}{p_3} = \frac{2}{n}\left(1 - \frac{1}{r_2} + \frac{1}{q}\right) - \frac{1}{n}\left(k - \frac{n}{p}\right) \tag{5.96}$$

hold. Here n is the dimension of the domain $(n=2,3)$, $k=0,1$, *and it is assumed that* $p, p_1, p_2, p_3, q > 1$ *and* $r_1, r_2 \geq 1, p, q \geq p_2$.

This theorem is proved in the same way as Theorem 4.2.

We next introduce the space $S_p^{(l)}$—the closure of the set of smooth solenoidal vectors vanishing on S in the metric of $W_p^{(l)}$. In particular, $S_p^{(0)} = S_p$ and $S_2^{(1)} = H_1$.

THEOREM 5.2. *The solution of problem* (5.38) *satisfies*

$$\|\overline{u}(t)\|_{S_p^{(1)}} \leq Ce^{-\sigma_0 t}\left\{\|\overline{u}_0\|_{S_p^{(1)}} + \left[\int_0^t e^{r\sigma,\tau}\|\overline{f}(\tau)\|_{S_{p_1}}^r \, d\tau\right]^{1/r}\right\}; \quad (5.97)$$

$$\left(\frac{n}{p} = \frac{n}{p_1} + 1 - \frac{2}{r'}; \quad n=2,3; \quad p, p_1 > 1; \quad r \geq 1\right).$$

PROOF. In the case $\overline{u}_0 = 0$ the theorem follows immediately from Lemma 2.6 and the analogue of Lemma 4.1. We therefore treat the case $\overline{f} = 0$.

We consider the operator A on $S_p^{(1)}$. We suppose that its domain is the set $D_p^{(1)}(A)$ of vectors $\overline{u} \in S_p^{(1)} \cap W_p^{(3)}$ such that $A\overline{u} \in S_p^{(1)}$. The operator is closed and generates an analytic semigroup in $S_p^{(1)}$. (This fact is deduced from Theorem 6.3 in the same way that Lemma 5.7 is deduced from Theorem 6.1.) Therefore, (5.97) holds for $\overline{f} = 0$. The theorem is proved.

THEOREM 5.3. *The solution of problem* (5.38) *satisfies*

$$\begin{aligned} &\int_0^t e^{r\sigma_0\tau}\left(\left\|\frac{\partial u}{\partial t}\right\|_{L_p(\Omega)}^r + \|\overline{u}\|_{W_p^{(2)}(\Omega)}^r\right) d\tau \\ &\qquad \leq C\left\{|||\overline{u}_0|||_p^r + \int_0^t e^{r\sigma_0\tau}\|\overline{f}\|_{L_p(\Omega)}^r \, d\tau\right\}. \end{aligned} \quad (5.98)$$

PROOF. For $r = p > 1$ this theorem is known (see [77], [78], and also §7). For any $r, p > 1$ it follows from Theorem 3.1. (See also the appendix to §5.)

Making Theorem 3.2 concrete, in the case of the Navier-Stokes equations we arrive at the following assertion.

THEOREM 5.4. *Suppose the homogeneous equation*

$$\frac{du}{dt} + Au = 0 \quad (5.99)$$

has no periodic solutions except the zero solution. Then the inhomogeneous equation

$$\frac{du}{dt} + Au = f(t) \tag{5.100}$$

for any T-periodic vector-valued function $f \in L_r((0,T), S_p)$ *has a unique periodic solution, and*

$$\int_0^T \left(\left\| \frac{\partial u}{\partial t} \right\|^r_{L_p(\Omega)} + \|u\|^r_{W_p^{(2)}(\Omega)} \right) dt \le C \int_0^T \|f\|^r_{S_p}\, dt. \tag{5.101}$$

Appendix to §5

In this appendix it is shown how coerciveness inequalities for the Cauchy problem on an infinite time interval can be derived from coerciveness inequalities on a finite interval.

We consider the Cauchy problem for an equation in a Banach space X

$$\dot{u} + Au = f; \qquad u(0) = a. \tag{1}$$

We assume that A is the generator of an analytic semigroup and that coerciveness holds on the segment $[0, T]$, i.e., for any $T > 0$

$$\begin{aligned} \|u\|^p_{B_p(0,T)} &\equiv \int_0^T (\|\dot{u}\|^p + \|Au\|^p)\, dt \\ &\le C_0^p \left[\int_0^T \|f(t)\|^p\, dt + |||a|||^p_p \right]. \end{aligned} \tag{2}$$

We shall show that then

$$\begin{aligned} \|u\|^p_{B_p(0,\tau;\sigma)} &\equiv \int_0^\tau e^{p\sigma t}(\|\dot{u}\|^p + \|Au\|^p)\, dt \\ &\le C^p \left[\int_0^\tau e^{p\sigma t} \|f\|^p\, dt + |||a|||^p_p \right] \end{aligned} \tag{3}$$

with a constant C not depending on τ $(0 < \tau \le \infty)$ for any σ such that $\operatorname{Re}\sigma(A) > \sigma$. For brevity we shall prove (3) for $\tau = \infty$.

Let W_p be the Banach space obtained by completing D_A in the norm

$$\|a\|_{W_p} = |||a|||_p + \|a\|_X. \tag{4}$$

We note that the operator A generates an analytic semigroup in W_p. Indeed, suppose $\Phi \in B_p(0, \tau)$ and $\Phi(0) = g$. Then $R_\lambda \Phi(0) = R_\lambda g$, and for any $\lambda \in \rho(A)$

$$|||R_\lambda g|||_p \le \|R_\lambda \cdot \Phi\|_{B_p(0,T)} \le \|R_\lambda\|_{X \to X} \|\Phi\|_{B_p(0,T)}. \tag{5}$$

Passing to the infimum over Φ, we find that for any $g \in W_p$

$$|||R_\lambda g|||_p \leq \|R_\lambda\|_{X\to X}|||g|||_p. \tag{6}$$

Our assertion now follows from a theorem of M. Z. Solomyak and the assumption that $-A$ is the generator of an analytic semigroup in X.

Thus, for the solution of the Cauchy problem (1) for $f = 0$ we have the estimate

$$|||u(t)|||_p \leq C_1 e^{-\sigma_1 t}|||a|||_p, \qquad t \geq 0 \tag{7}$$

for any $\sigma_1 < \operatorname{Re}\sigma(A)$.

On the segment $[nT, (n+1)T] = I_n$, $n = 0, 1, 2, \dots$, we represent the solution of problem (1) in the form

$$u = w_n + v_n; \qquad w_n = \sum_{k=0}^{n} u_k. \tag{8}$$

We define the vector-valued functions u_k and v_k for $t \geq kT$ as solutions of the Cauchy problems

$$\begin{aligned} \dot v_k + Av_k = f; \qquad & v_k(kT) = 0; \\ \dot u_k + Au_k = 0; \qquad & u_k(kT) = v_{k-1}(kT). \end{aligned} \tag{9}$$

For convenience we introduce the notation $v_{-1}(0) = a$.

Applying inequality (2) to (1) on the segments I_n and summing, we obtain

$$\|u\|^p_{B_p(0,\infty;\sigma)} \leq C_2^p \left\{ \int_0^\infty e^{p\sigma t}\|f(t)\|^p\,dt + \sum_{n=0}^{\infty} e^{p\sigma nT}|||w_n(nT)|||_p^p \right\},$$

$$C_2 = 2^{1/p'}Ce^{\sigma T}. \tag{10}$$

We estimate the last term in (10).

Suppose σ_0 satisfies the condition $\sigma < \sigma_0 < \operatorname{Re}\sigma(A)$ and the positive δ is so small that $\sigma_1 = \sigma_0 + \delta$ satifies the same condition. Applying (7), (8), and the Hölder inequality for sums, we obtain

$$|||w_n(nT)|||^p \leq C_1^p \sum_{k=0}^{n} e^{-p\sigma_0(n-k)T}|||v_{k-1}(kT)|||_p^p \times \left(\sum_{k=0}^{n} e^{-\delta p'(n-k)T}\right)^{p-1}. \tag{11}$$

From (11) we conclude that

$$|||w_n(nT)|||^p \leq C_3^p \sum_{k=0}^{n} e^{-p\sigma_0(n-k)T}|||v_{k-1}(kT)|||^p,$$

$$C_3 = C_1 e^{\delta T}(e^{\delta p' T} - 1)^{-1/p}. \tag{12}$$

Using (12) to estimate the sum in (10), we obtain, after a change of the order of summation,

$$\sum_{n=0}^{\infty} e^{p\sigma nT} |||w_n(nT)|||_p^p \leq C_3^p \sum_{k=0}^{\infty} e^{p\sigma_0 kT} |||v_{k-1}(kT)|||_p^p \times \sum_{n=k}^{\infty} e^{p(\sigma-\sigma_0)nT} = C_4^p \sum_{k=0}^{\infty} e^{p\sigma kT} |||v_{k-1}(kT)|||_p^p; \tag{13}$$

$$C_4 = C_3(1 - e^{-p(\sigma_0-\sigma)T})^{-1/p}.$$

From (13) we obtain

$$\sum_{n=0}^{\infty} e^{p\sigma nT} |||w_n(nT)|||_p^p \leq C_4^p \left[\sum_{k=1}^{\infty} e^{p\sigma kT} \|v_{k-1}\|_{B_p(I_{k-1})}^p + |||a|||_p^p\right]. \tag{14}$$

Further, using inequality (2) applied to problem (9), from (14) we conclude that

$$\sum_{n=0}^{\infty} e^{p\sigma nT} |||w_n(nT)|||_p^p \leq C_4^p \left[|||a|||_p^p + \sum_{k=1}^{\infty} e^{p\sigma kT} \int_{(k-1)T}^{kT} \|f(t)\|^p \, dt\right] \leq C_4^p \left[|||a|||_p^p + e^{p\sigma T} \int_0^{\infty} e^{p\sigma t} \|f(t)\|^p \, dt\right]. \tag{15}$$

From (10) and (15) we obtain

$$\|u\|_{B_p(0,\infty;\sigma)}^p \leq C_2^p \left\{(1 + C_4^p e^{p\sigma T}) \int_0^{\infty} e^{p\sigma t} \|f\|^p \, dt + C_4^p |||a|||_p^p\right\}, \tag{16}$$

which essentially coincides with (3).

We now show that under the hypotheses of Theorem 3.2 for T-periodic solutions it is possible to choose the constant in the coerciveness inequality (3.5) not to depend on T. We shall assume that $\operatorname{Re} \sigma(A) > 0$ (in the general case it is further necessary to make an estimate for the part A_- of the operator A corresponding to the subset of the spectrum lying in the left half-plane; this is done similarly and even more simply, since A_- is a bounded operator). Inequality (3) for $\sigma = 0$ and $\tau = nT$ for a periodic solution u_0 gives us

$$\int_0^{nT} [\|\dot{u}_0\|^p + \|Au_0\|^p] \, dt \leq C^p \left[\int_0^{nT} \|f\|^p \, dt + |||u(0)|||_p^p\right] \tag{17}$$

with a constant C depending on p but not on f, n, or T. We rewrite (17) in the form

$$n\int_0^T (\|\dot{u}\|^p + \|Au\|^p)\,dt \le C^p\left[n\int_0^T \|f\|^p\,dt + |||u(0)|||_p^p\right]. \tag{18}$$

From (18) for $n \to \infty$ we obtain finally

$$\int_0^T (\|\dot{u}\|^p + \|Au\|^p)\,dt \le C^p\int_0^T \|f\|^p\,dt. \tag{19}$$

§6. An estimate of the resolvent of the linearized Navier-Stokes operator

In a bounded three-dimensional domain Ω with boundary S we consider the boundary value problem

$$\sigma\overline{u} - \Delta\overline{u} = -\nabla P + \overline{f}; \tag{6.1}$$

$$\operatorname{div}\overline{u} = 0; \tag{6.2}$$

$$\overline{u}/S = 0. \tag{6.3}$$

In order to eliminate the ambiguity in the definition of the pressure we suppose, for example, that $\int_\Omega P\,dx = 0$.

Estimates of the solution $\overline{u}, P$ of the given problem and of its derivatives in L_p are given in this section. The main results are formulated in Theorems 6.1–6.3.

THEOREM 6.1. *Suppose $S \in C^2$ and $\overline{f} \in L_p(\Omega)$ $(p > 1)$. Then the boundary value problem* (6.1)–(6.3) *has a unique solution $\overline{u} \in W_p^{(2)}, P \in W_p^{(1)}$, and*

$$\|\overline{u}\|_{W_p^{(2)}} \le C\|\overline{f}\|_{L_p}; \tag{6.4}$$

$$\|\overline{u}\|_{L_p} \le \frac{C}{1+|\sigma|}\|\overline{f}\|_{L_p}; \tag{6.5}$$

$$\|\nabla P\|_{L_p} \le C\|\overline{f}\|_{L_p} \tag{6.6}$$

for all σ in the sector $\Sigma_{0,\alpha} = \{\sigma : \pi \ge |\arg(-\sigma)| \ge \alpha\}$, $0 < \alpha < \pi/2$, and the constant C depends only on Ω, p, and α, but not on $\overline{f}$.

We first prove several lemmas; the center of gravity of the proof of Theorem 6.1 lies on Lemma 6.3.

We begin from a problem of vector analysis. Suppose $\overline{u} = \overline{u}(x)$ is a smooth vector which is solenoidal in the domain Ω and has flux 0 across the boundary S. We pose the problem of determining a vector $\overline{B}$ from the conditions

$$\operatorname{div}\overline{B} = 0; \tag{6.7}$$

$$\operatorname{curl}\overline{B} = \overline{u}; \tag{6.8}$$

$$B_n/S = 0. \tag{6.9}$$

It is well known that if the domain Ω is not simply connected, say, $(m+1)$-connected, then a solution of this problem is determined not uniquely but only up to a harmonic term. The set of harmonic vectors forms a finite-dimensional space; as basis it is possible to take, for example, vectors $\overline{\psi}_k$ satisfying the conditions

$$\operatorname{div}\overline{\psi}_k = 0; \quad \operatorname{curl}\overline{\psi}_k = 0; \quad \psi_k \cdot \overline{n}/S = 0, \quad \int_{\gamma_l} \overline{\psi}_k \cdot d\overline{x} = \delta_{kl}$$
$$(k, l = 1, 2, \dots, m),$$

where $\gamma_1, \dots, \gamma_m$ is a complete collection of independent one-dimensional cycles ("a one-dimensional homology basis"). Uniqueness of a solution can then be achieved by imposing the additional condition

$$\int_{\gamma_k} \overline{B} \cdot d\overline{x} = C_k; \qquad (k = 1, 2, \dots, m), \tag{6.10}$$

where C_k are known constants.

LEMMA 6.1. *Suppose* $\overline{u} \in S_p$ $(p > 1)$. *Then there exists a vector* $\overline{B} \in W_p^{(1)}$ *satisfying* (6.7)–(6.9) *and*

$$\|\overline{B}\|_{W_p^{(1)}} \leq C\|\overline{u}\|_{L_p}, \tag{6.11}$$

where the constant C *depends only on* Ω *and* p, *not on* $\overline{u}$.

PROOF. It suffices to carry out the arguments in the case of a smooth solenoidal vector $\overline{u}$ vanishing in a boundary strip. (The set of such vectors is dense in S_p; therefore, using the fact that generalized differentiation is closed and (6.11), we go over immediately from the special case to the general case.) We extend the vector $\overline{u}(x)$ to all of R^3 by setting it equal to 0 outside Ω, and define a vector $\overline{B}_0$ as a solution of the problem in the whole space

$$\operatorname{div}\overline{B}_0 = 0; \quad \operatorname{curl}\overline{B}_0 = \overline{u}; \quad \overline{B}_0/_\infty = 0. \tag{6.12}$$

A solution of problem (6.12) has the form

$$\overline{B}_0 = \operatorname{curl}\overline{C}_0,$$

where the vector $\overline{C}_0$ is a solution of the Poisson equation

$$\Delta\overline{C}_0 = -\overline{u}; \qquad \overline{C}_0/_\infty = 0. \tag{6.13}$$

By a theorem of Koshelev [33], from (6.13) we obtain

$$\|\overline{B}_0\|_{W_p^{(1)}(\Omega)} \leq C_1\|\overline{C}_0\|_{W_p^{(2)}(\Omega)} \leq C\|\overline{u}\|_{L_p}. \tag{6.14}$$

It is clear that all the conditions (6.7)–(6.9) are satisfied if we set $\overline{B} = \overline{B}_0 + \operatorname{grad}\varphi$, where φ is the solution of the Neumann problem

$$\Delta\varphi = 0; \qquad \left.\frac{\partial\varphi}{\partial n}\right|_S = -B_{0n}. \tag{6.15}$$

The condition for solvability of this problem is satisfied by the first of equations (6.12). We further note that B_{0n} is the boundary value of some function in $W_p^{(1)}$. Indeed, let $n_k(x)$ be a function defined in the domain Ω and equal on the boundary S to the kth coordinate of the vector of the outer normal $\overline{n}$. Since $S \in C^2$, the function $n_k(x)$ on S has smoothness C^1, and hence, as is known ([54], §16), we can assume that it has smoothness C^1 also in Ω. We thus have

$$B_{0n} = B_{0k}(x)n_k(x)|_S; \qquad \|B_{0k}n_k\|_{W_p^{(1)}} \leq C\|\overline{B}_0\|_{W_p^{(1)}}. \tag{6.16}$$

Using a known result of [1] and inequalities (6.14) and (6.16), from (6.15) we derive

$$\|\varphi\|_{W_p^{(2)}} \leq C\|B_{0k}n_k\|_{W_p^{(1)}}, \tag{6.17}$$

which together with (6.14) leads to (6.11). The lemma is proved.

LEMMA 6.2. *If in addition to the conditions of Lemma* 6.1 *it is known that* $S \in C^{l+1}$ *and* $\overline{u} \in W_p^{(l)}$ $(p > 1, l \geq 0)$, *then*

$$\|B\|_{W_p^{(l+1)}} \leq C\|\overline{u}\|_{W_p^{(l)}}. \tag{6.18}$$

PROOF. The proof proceeds by induction on l. For $l = 0$ the required result has already been established in Lemma 6.1. Suppose now that the lemma is true for some $l \geq 0$. We show that it is also true for $l + 1$. Let x_0 be an interior point of Ω such that its distance from the boundary S is not less than $3h$. We introduce the function $\chi = \chi_h(x)$ by setting

$$\chi_h(x) = \psi\left(\frac{|x - x_0|}{2h}\right), \tag{6.19}$$

where ψ is the function defined in (5.25). We define the vector $\overline{B}' = \chi\overline{B}$. Clearly $\overline{B}' = \overline{B}$ in an h-neighborhood $\Omega_{x_0,h}$ of the point x_0, and we can assume that it is defined on the whole space R^3 by setting $\overline{B}' = 0$ outside

$\Omega_{x_0,2h}$. From (6.7) and (6.8) it follows that the vector $\overline{B}'$ in all of R^3 satisfies

$$\operatorname{div}\overline{B}' = \nabla\chi\cdot\overline{B}; \qquad \operatorname{curl}\overline{B}' = \chi\overline{u} + \nabla\chi\times\overline{B}. \tag{6.20}$$

Hence $\overline{B}'$ admits the representation

$$\overline{B}' = \operatorname{grad}\theta' + \operatorname{curl}\overline{C}', \tag{6.21}$$

where θ' and $\overline{C}'$ are defined as solutions vanishing at infinity of the Poisson equations

$$\Delta\theta' = \nabla\chi\cdot\overline{B}; \qquad -\Delta\overline{C}' = \chi\overline{u} + \nabla\chi\times\overline{B}. \tag{6.22}$$

From (6.21) and (6.22) we derive

$$\|D^{l+1}\overline{B}\|_{L_p(\Omega_{x_0,h})} \le C(\|\overline{B}\|_{W_p^{(l)}(\Omega_{x_0,2h})} + \|\overline{u}\|_{W_p^{(l)}}) \tag{6.23}$$

We now need to obtain an analogous estimate in the case $x_0\in S$. We choose the origin at x_0 and direct the x_3 axis along the inner normal to S. Suppose the equation of the boundary S near $x_0 = 0$ is $x_3 = \gamma(x_1, x_2)$. By hypothesis $\gamma\in C^{l+1}$ in some neighborhood $\Omega_{x_0,h}$ of x_0. In this neighborhood we introduce new coordinates:

$$\alpha_1 = x_1; \qquad \alpha_2 = x_2; \qquad \alpha_3 = x_3 - \gamma(x_1, x_2). \tag{6.24}$$

The transformation (6.24) is one-to-one and takes a portion of the boundary S in the neighborhood of x_0 into some region on the plane $\alpha_3 = 0$. From (6.7)–(6.9) we deduce that

$$\operatorname{div}_\alpha\overline{B} = \gamma_{x_k}\frac{\partial B_k}{\partial\alpha_3}; \tag{6.25}$$

$$\operatorname{curl}_\alpha\overline{B} = \overline{u} + \overline{q}; \tag{6.26}$$

$$B_3|_{\alpha_3=0} = \gamma_{x_k}B_k, \tag{6.27}$$

where the vector $\overline{q} = \overline{q}(\overline{B})$ has coordinates

$$q_1 = \gamma_{x_2}\frac{\partial B_3}{\partial\alpha_3}; \quad q_2 = -\gamma_{x_1}\frac{\partial B_3}{\partial\alpha_3}; \quad q_3 = \gamma_{x_1}\frac{\partial B_2}{\partial\alpha_3} = \gamma_{x_2}\frac{\partial B_1}{\partial\alpha_3}. \tag{6.28}$$

We now introduce the vector $\hat{\overline{B}}$ by setting

$$\hat{\overline{B}} = \chi(\alpha)\overline{B}; \quad \chi(\alpha) = \psi\left(\frac{|\alpha|}{2h}\right). \tag{6.29}$$

By (6.25)–(6.27) the vector $\hat{\overline{B}}$ in the half-space $\alpha_3\ge 0$ satisfies

$$\operatorname{div}_\alpha\hat{\overline{B}} = \gamma_{x_k}\frac{\partial\hat{B}_k}{\partial\alpha_3} + \left(\chi_{\alpha_k} - \gamma_{x_k}\frac{\partial\chi}{\partial\alpha_3}\right)B_k; \tag{6.30}$$

$$\operatorname{curl}_\alpha\hat{\overline{B}} = \overline{q}(\hat{\overline{B}}) + \chi\overline{u} + \overline{q}_0 + \nabla_\alpha\chi\times B; \tag{6.31}$$

$$\hat{B}_3|_{\alpha_3=0} = \gamma_{x_k}\hat{B}_k. \tag{6.32}$$

Here outside a $2h$-neighborhood of $\alpha = 0$ we set $\hat{\overline{B}} = 0$. The vector $\overline{q}_0$ is determined by

$$q_{01} = -\gamma_{x_2} B_3 \frac{\partial \chi}{\partial \alpha_3}; \qquad q_{02} = \gamma_{x_1} \frac{\partial \chi}{\partial \alpha_3};$$
$$q_{03} = (B_1 \gamma_{x_2} - B_2 \gamma_{x_1}) \frac{\partial \chi}{\partial \alpha_3}. \tag{6.33}$$

In the half-space $\alpha_3 > 0$ we now consider the boundary value problem

$$\operatorname{div} \overline{E} = \rho; \quad \operatorname{curl} \overline{E} = \overline{g}; \quad E_3|_{\alpha=0} = \pi(\alpha)|_{\alpha_3=0}, \tag{6.34}$$

where $\rho(\alpha), \overline{g}(\alpha)$, and $\pi(\alpha)$ are defined everywhere in the half-space $\alpha_3 \geq 0$, are compactly supported, and have smoothness $W_p^{(l)}$, $p > 1$. We shall demonstrate the estimate

$$\|D_\alpha^{l+1} \overline{E}\|_{L_p} \leq C(\|\rho\|_{W_p^{(l)}} + \|\overline{g}\|_{W_p^{(l)}} + \|\pi\|_{W_p^{(l)}}), \tag{6.35}$$

where C depends only on p.

Indeed, the vector $\overline{E}$ admits the representation

$$\overline{E} = \operatorname{grad} \theta + \overline{E}_0, \tag{6.36}$$

where the function θ and the components of the vector $\overline{E}_0$ are determined by solving the Dirichlet and Neumann problems in the half-space $\alpha_3 > 0$:

$$\Delta \theta = \rho; \qquad \left.\frac{\partial \theta}{\partial \alpha_3}\right|_{\alpha_3=0} = \pi(\alpha)\Big|_{\alpha_3=0}; \tag{6.37}$$

$$\Delta \overline{E}_0 = \operatorname{curl} \overline{g}; \tag{6.38}$$

$$E_{03} = 0; \qquad \frac{\partial E_{01}}{\partial \alpha_3} = g_2 + \frac{\partial E_{03}}{\partial \alpha_1};$$
$$\frac{\partial E_{02}}{\partial \alpha_3} = -g_1 + \frac{\partial E_{03}}{\partial \alpha_2}; \qquad (\alpha_3 = 0). \tag{6.39}$$

Therefore, known estimates in L_p of solutions of these problems (see [1] and [33]) immediately lead us to (6.35); we remark only that the components of the vector $\overline{E}_0$ can be estimated successively, beginning with E_{03}.

To derive the estimate of $\hat{\overline{B}}$ from (6.30)–(6.32) we now apply (6.35). We note that

$$D_\alpha^l \left(\gamma_{x_k} \frac{\partial \hat{B}_k}{\partial \alpha_3} \right) = \gamma_{x_k} \frac{\partial D^l \hat{B}_k}{\partial \alpha_3} + \cdots; \tag{6.40}$$

$$D_\alpha^l \overline{q} \left(\hat{\overline{B}} \right) = \overline{q} \left(D_\alpha^l \hat{\overline{B}} \right) + \cdots,$$

where the dots denote expressions containing derivatives of the vector $\overline{B}$ of order no higher than l and derivatives of the function γ of order no higher than $l+1$. Similarly,

$$D^l(\gamma_{x_k}, \hat{B}_k) = \gamma_{x_k} D^l \hat{B}_k + \cdots, \tag{6.41}$$

where terms have been dropped which contain derivatives of $\overline{B}$ of order no higher than $l-1$ and derivatives of γ of order no higher than $l+1$. As a result we have

$$\|D^{l+1}\hat{\overline{B}}\|_{L_p} \leq C\left[\max_{\alpha\in\Omega 0,2h} |\nabla\gamma| \cdot \|D^{l+1}\hat{\overline{B}}\|_{L_p} + \|\overline{B}\|_{W_p^{(l)}(\Omega_{0,2h})} + \|\overline{u}\|_{W_p^{(l)}(\Omega_{0,2h})}\right]. \tag{6.42}$$

We now choose h so small that

$$C \max_{\alpha\in\Omega_{0,2h}} |\nabla\gamma| < \frac{1}{2}. \tag{6.43}$$

This is possible, since $\nabla\gamma = 0$ for $\alpha = 0$. From (6.42), recalling that $\hat{\overline{B}} = \overline{B}$ for $|\alpha| \leq h$ and returning from the variables α to the variables x, we then obtain (possibly with another constant h)

$$\|D^{l+1}\overline{B}\|_{L_p(\Omega_{x_0,h})} \leq C\left[\|u\|_{W_p^{(l)}(\Omega_{x_0,2h})} + \|\overline{B}\|_{S_p^{(l)}(\Omega_{x_0,2h})}\right]. \tag{6.44}$$

From (6.23) and (6.24) with the help of the usual argument using the Borel lemma we deduce that

$$\|\overline{B}\|_{W_p^{(l+1)}(\Omega)} \leq C(\|\overline{u}\|_{W_p^{(l)}(\Omega)} + \|\overline{B}\|_{W_p^{(l)}(\Omega)}). \tag{6.45}$$

Now (6.18) follows directly from (6.45) and the induction hypothesis. Lemma 6.2 is proved.

LEMMA 6.3. *Suppose*

$$\sigma \in \Sigma_{0,\theta} = \{\sigma\colon \theta \leq |\arg(-\sigma)| \leq \pi\}, \qquad 0 < \theta < \pi/2,$$

while the vector $\overline{u} = \overline{u}(x, z)$ $(x = (x_1, x_2), z = x_3)$ *together with the function* $P(x, z)$ *forms a solution of the boundary value problem*

$$\sigma\overline{u} - \Delta\overline{u} = -\nabla P + \overline{f}; \tag{6.46}$$

$$\operatorname{div} \overline{u} = g = \sum_{k=1}^{3} \frac{\partial g_k}{\partial x_k}; \tag{6.47}$$

$$\overline{u}/_{z=0} = 0, \tag{6.48}$$

in the half-space $z > 0$, *where* $\overline{f}, g$, *and* g_k *are known. Then*

$$|\sigma| \cdot \|\overline{u}\|_{L_p} + \|\overline{u}\|_{W_p^{(2)}} + \|\nabla P\|_{L_p} \leq C\left(\|\overline{f}\|_{L_p} + \|g\|_{W_p^{(1)}} + |\sigma| \sum_{k=1}^{3} \|g_k\|_{L_p}\right), \tag{6.49}$$

where C *depends only on* θ, *and* $p > 1$.

PROOF. We represent the vector $\overline{u}$ in the form

$$\overline{u} = \overline{u}_0 + \operatorname{grad}\varphi + \overline{w}, \tag{6.50}$$

where $\overline{u}_0$, φ, and $\overline{w}$ are determined by solving the following boundary value problems in the half-space $z > 0$:

$$\sigma\overline{u}_0 - \Delta\overline{u}_0 = \overline{f}; \quad \overline{u}_0/_{z=0} = 0; \quad \overline{u}_0/_{\infty} = 0; \tag{6.51}$$

$$\begin{gathered} \Delta\varphi = g - \operatorname{div}\overline{u}_0 = \sum_{k=1}^{3} \frac{\partial}{\partial x_K}(g_k - u_{0k}); \\ \left.\frac{\partial\varphi}{\partial z}\right|_{z=0} = 0; \qquad \varphi|_{\infty} = 0; \end{gathered} \tag{6.52}$$

$$\begin{gathered} \overline{w} = \operatorname{curl}\overline{G}; \qquad \overline{G} = (-\psi_{x_2}, \psi_{x_1}, 0); \\ \Delta^2\psi - \sigma\Delta\psi = 0; \quad \psi|_{z=0} = 0; \quad \left|\frac{\partial\psi}{\partial z}\right|_{z=0} = \varphi(x, 0). \end{gathered} \tag{6.53}$$

By results of [74] (see also Theorem 4.1) the vector $\overline{u}_0$ admits the estimate

$$|\sigma| \cdot \|\overline{u}_0\|_{L_p} + \|\overline{u}_0\|_{W_p^{(2)}} \leq C\|\overline{f}\|_{L_p}. \tag{6.54}$$

By applying Theorem 1.6 and taking note of (6.54), for the function φ we obtain

$$\|D_x^3\varphi\|_{L_p} \leq C(\|g\|_{W_p^{(1)}} + \|\overline{f}\|_{L_p}); \tag{6.55}$$

$$\begin{aligned} |\sigma| \cdot \|\nabla\varphi\|_{L_p} &\leq |\sigma| C \sum_{k=1}^{3} \|g_k - u_{0k}\|_{L_p} \\ &\leq C\left(|\sigma| \sum_{k=1}^{3} \|g_k\|_{L_p} + \|\overline{f}\|_{L_p}\right). \end{aligned} \tag{6.56}$$

We now take up problem (6.53). Applying separation of variables, we write its solution in the form

$$\psi(x, z) = \int_{R^2} e^{i(x,y)} \frac{e^{\lambda_1 z} - e^{\lambda_2 z}}{\lambda_1 - \lambda_2} \hat{\varphi}(y, 0)\, dy, \tag{6.57}$$

where $\hat{\varphi}$ is the Fourier transform of $\varphi(x, 0)$,

$$\hat{\varphi}(y, 0) = \frac{1}{4\pi^2} \int_{R^2} \varphi(x, 0) e^{-i(x,y)}\, dx, \tag{6.58}$$

and λ_1 and λ_2 are given by

$$\lambda_1 = -|y| = -\sqrt{y_1^2 + y_2^2}; \qquad \lambda = -\sqrt{|y|^2 + \sigma}. \tag{6.59}$$

example, it suffices to assume that $\overline{u}$ is a weak solution in the sense that $\overline{u} \in L_p$ and

$$\int_{z>0} \overline{u}(\sigma\overline{\Phi} - \Delta\overline{\Phi})\,dx\,dz = \int_{z>0} \overline{f}\cdot\overline{\Phi}\,dx\,dz; \tag{6.86}$$

$$\int_{z>0} \nabla\theta\cdot\overline{u}\,dx\,dz = -\int_{z>0} g\theta\,dx\,dz; \tag{6.87}$$

for arbitrary compactly supported, infinitely differentiable $\overline{\Phi}$ and θ, where $\operatorname{div}\overline{\Phi} = 0$ and $\overline{\Phi}|_{z=0} = 0$. We shall not prove this here; the proof is carried out according to familiar models (see, for example, [1], [82], or [78]); actually, of course, the matter consists in the fact that for smooth $\overline{f}$ and g the equalities from which (6.49) is derived (for example, (6.57), etc.) follow directly from (6.86) and (6.87) without using (6.46)–(6.48).

PROOF OF THEOREM 6.1. We begin from the derivation of the a priori estimate (6.49) under the assumption that $\overline{u} \in W_p^{(2)}$ and $P \in W_p^{(1)}$ satisfy conditions (6.1)–(6.3).

Let $x_0 \in S$ be an arbitrary boundary point. We assume that in a neighborhood of x_0 the surface S in a local coordinate system is given by

$$x_3 = \gamma(x_1, x_2), \tag{6.88}$$

where $\gamma \in C^2$, and the x_3-axis is directed along the inner normal. We make the change of variables

$$\alpha_1 = x_1; \qquad \alpha_2 = x_2; \qquad \alpha_3 = x_3 - \gamma(x_1, x_2). \tag{6.89}$$

The transformation (6.89) is obviously one-to-one and takes a portion of the surface S in a neighborhood of x_0 into some region on the plane $\alpha_3 = 0$. In the new coordinates (6.1) and (6.2) take the form

$$\begin{aligned}\sigma\overline{u} - \Delta_\alpha\overline{u} = {} & -\nabla_\alpha P + \frac{\partial P}{\partial\alpha_3}\gamma_{x_k}\overline{i}_k - 2\frac{\partial^2\overline{u}}{\partial\alpha_3\partial\alpha_k}\gamma_{x_k} \\ & + \frac{\partial^2\overline{u}}{\partial\alpha_3^2}(\nabla\gamma)^2 - \frac{\partial\overline{u}}{\partial\alpha_3}\Delta\gamma + \overline{f};\end{aligned} \tag{6.90}$$

$$\operatorname{div}_\alpha\overline{u} = \frac{\partial u_k}{\partial\alpha_3}\gamma_{x_k}, \tag{6.91}$$

where $\overline{i}_k$ is the unit vector of the kth coordinate.

Using the function defined in (6.29), we now introduce the new unknowns

$$\overline{u}^0 = \chi\overline{u}; \qquad P^0 = \chi P. \tag{6.92}$$

The quantity h must be so small that the support $\Omega_{0,2h}$ of the function χ is contained within the domain where the transformation (6.89) is defined. By (6.90) and (6.91), $\overline{u}^0$ and P^0 satisfy

$$\sigma u^0 - \Delta_\alpha \overline{u}^0 = -2\frac{\partial^2 \overline{u}^0}{\partial \alpha_k \partial \alpha_3}\gamma_{x_k} + \frac{\partial^2 \overline{u}^0}{\partial \alpha_3^2}(\nabla\gamma)^2 + \frac{\partial P^0}{\partial \alpha^3}\gamma_{x_k} i_k - \nabla_\alpha P^0 + \overline{f}^0. \tag{6.93}$$

$$\operatorname{div}_\alpha \overline{u}^0 = \frac{\partial u_k^0}{\partial \alpha_3}\gamma_{x_k} + g^0 \equiv g. \tag{6.94}$$

$$\overline{u}_0/s = 0, \tag{6.95}$$

where $\overline{f}^0$ and g^0 contain only lower order derivatives of the solution:

$$\begin{aligned}\overline{f}^0 = P\left(\nabla_\alpha \chi - \frac{\partial \chi}{\partial \alpha_3}\gamma_{x_k}\overline{i}_k\right) - 2\frac{\partial \overline{u}}{\partial \alpha_k}\left(\frac{\partial \chi}{\partial \alpha_k} - \frac{\partial \chi}{\partial \alpha_3}\gamma_{x_k}\right)\\ + \frac{\partial \overline{u}}{\partial \alpha_3}\left[2\frac{\partial \chi}{\partial \alpha_k}\gamma_{x_k} - 2\frac{\partial \chi}{\partial \alpha_3}(\nabla\gamma)^2 - \chi\Delta\gamma\right]\\ + \overline{u}\cdot\left[2\frac{\partial^2 \chi}{\partial \alpha_3 \partial \alpha_k}\gamma_{x_k} - \frac{\partial^2 \chi}{\partial \alpha_3^2}(\nabla\gamma)^2 - \Delta_\alpha \chi\right] + \chi\overline{f};\end{aligned} \tag{6.96}$$

$$g^0 = u_k\left(\frac{\partial \chi}{\partial \alpha_k} - \frac{\partial \chi}{\partial \alpha_3}\gamma_{x_k}\right). \tag{6.97}$$

Introducing the "vectorial flux function" $\overline{B}$ according to (6.7)–(6.9), we represent the function g in the form

$$g(\alpha) = \sum_{k=1}^{3}\frac{\partial g_k}{\partial \alpha_k}; \tag{6.98}$$

$$\begin{aligned} g_k = [\overline{B}\times\nabla_x\chi]_k \qquad (k = 1, 2);\\ g_3 = [\overline{B}\times\nabla_x\chi]_3 + \nabla\gamma\cdot(\overline{u}^0 - \overline{B}\times\nabla_x\chi).\end{aligned} \tag{6.99}$$

Equations (6.93) and (6.94) may be assumed satisfied in the entire half-space $\alpha_3 > 0$ if we assume that $\overline{u}^0, P^0, \overline{f}^0$, and $\overline{g}^0$ vanish outside the ball $\Omega_{0,2h}$. We remark that $|\nabla\gamma| < \varepsilon$, ($\varepsilon > 0$ is any given number) in $\Omega_{0,2h}$ if h is sufficiently small. Therefore, applying Lemma 6.1 to problem (6.93)–(6.95), we arrive at the estimate

$$\begin{aligned}|\sigma|\cdot\|\overline{u}^0\|_{L_p} + \|\overline{u}^0\|_{W_p^{(2)}} + \|\nabla P^0\|_{L_p}\\ \le C(\|\overline{f}\|_{L_p(\Omega_0^0,2h)} + \varepsilon\|\overline{u}_0\|_{W_p^{(2)}}\\ + \|\overline{u}\|_{W_p^{(1)}(\Omega_0^0,2h)} + \|P\|_{L_p(\Omega_0^0,2h)}\\ + \varepsilon\|\nabla P^0\|_{L_p} + \varepsilon|\sigma|\|\overline{u}^0\|_{L_p} + |\sigma|\|\overline{B}\|_{L_p(\Omega_0^0,2h)}.\end{aligned} \tag{6.100}$$

Assuming ε sufficiently small (say, $\varepsilon < 1/2C$), from (6.100) we obtain

$$|\sigma| \cdot \|\overline{u}^0\|_{L_p} + \|\overline{u}^0\|_{W_p^{(2)}} + \|\nabla P^0\|_{L_p} \leq C(\|\overline{f}\|_{L_p(\Omega_0^0,2h)} + \|\overline{u}\|_{W_p^{(1)}(\Omega_0^0,2h)} + \|P\|_{L_p(\Omega_0^0,2h)} + |\sigma| \cdot \|\overline{B}\|_{L_p(\Omega_0^0,2h)}. \tag{6.101}$$

The time has arrived to recall that $\overline{u}^0 = \overline{u}$ for $\alpha \in \Omega_{0,h}^0$. Returning to the old variables, from (6.101) we derive the estimate (with a changed h)

$$|\sigma|\|\overline{u}\|_{L_p(\Omega_{x_0,h})} + \|\overline{u}\|_{W_p^{(2)}}(\Omega_{x_0,h}) + \|\nabla P\|_{L_p(\Omega_{x_0,h})} \leq C(\|\overline{f}\|_{L_p(\Omega_{x_0,2h})} + \|\overline{u}\|_{W_p^{(1)}(\Omega_{x_0,2h})} + \|P\|_{L_p(\Omega_{x_0,2h})} + |\sigma|\|\overline{B}\|_{L_p(\Omega_{x_0,2h})}) \tag{6.102}$$

We choose a finite collection of points $x_0^1, \dots, x_0^k$ such that the boundary strip Ω_h of width h is covered by the sets $\Omega_{x_0^r,h}$ $(r = 1, \dots, k)$. Writing (6.102) for each of these points, raising these inequalities to the power p, and summing, we obtain

$$|\sigma|\|\overline{u}\|_{L_p(\Omega_h)} + \|\overline{u}\|_{W_p^{(2)}(\Omega_h)} + \|\nabla P\|_{L_p(\Omega_h)} \leq C(\|\overline{f}\|_{L_p(\Omega_{2h})} + \|\overline{u}\|_{W_p^{(1)}(\Omega_{2h})} + \|P\|_{L_p(\Omega_{2h})} + |\sigma|\|\overline{B}\|_{L_p(\Omega_{2h})}). \tag{6.103}$$

It is considerably easier to derive an analogous estimate for the interior subdomain $\Omega - \Omega_h$. Indeed, let η be an infinitely differentiable function such that

$$\eta(x) = \begin{cases} 1, & x \in \Omega - \Omega_h; \\ 0, & x \in \Omega_{h/2}. \end{cases} \tag{6.104}$$

We introduce the vector $\overline{u}'$ and the function P' by the equalities

$$\overline{u}' = \eta\overline{u}; \qquad P' = \eta P. \tag{6.105}$$

Then from (6.1)–(6.2) we deduce

$$\sigma\overline{u}' - \Delta\overline{u}' = -\nabla P' + \overline{f}' \tag{6.106}$$

$$\operatorname{div}\overline{u}' = g', \tag{6.107}$$

where $\overline{f}'$ and $\overline{g}'$ are given by

$$\overline{f}' = \eta\overline{f} + \nabla\eta \cdot P - 2\frac{\partial\eta}{\partial x_k} \cdot \frac{\partial\overline{u}}{\partial x_k} - \Delta\eta \cdot \overline{u}; \tag{6.108}$$

$$\overline{g}' = \nabla\chi \cdot \overline{u} = \sum_{k=1}^{3} \frac{\partial g_k'}{\partial x_k}; \qquad g_k' = [\overline{B} \times \nabla\chi]_k. \tag{6.109}$$

Equations (6.106) and (6.107) may be assumed satisfied for all $x \in R^3$ if $\overline{u}', P', \overline{f}'$, and $\overline{g}'$ are extended by zero to the complement of the region

$\Omega - \Omega_{h/2}$. There remains the sole "hard problem"—choosing one of the possible methods of deriving the estimate. Indeed, it is possible to represent the solution of problem (6.106), (6.107) as a Fourier integral and apply Mikhlin's theorem on multipliers; by extending the unknown functions periodically, it is possible to obtain a solution in the form of a Fourier series and use the Marcinkiewicz theorem; it is possible to represent the solution as an integral operator with a known kernel—the fundamental solution of the system (6.106), (6.107)—and apply the Calderón-Zygmund theorem on singular integrals; finally, and what we shall now do, it is possible to reduce (6.106) and (6.107) to known equations. Indeed, the solution of system (6.106), (6.107) admits the representation

$$\overline{u}' = \overline{u}'_0 + \operatorname{grad}\varphi'; \qquad P' = g' - \operatorname{div}\overline{u}'_0 - \sigma\varphi', \tag{6.110}$$

where the vector $\overline{u}'_0$ and the function φ' are solutions vanishing at infinity of the equations

$$\sigma\overline{u}'_0 - \Delta\overline{u}'_0 = \overline{f}' \tag{6.111}$$

$$\Delta\varphi' = g' - \operatorname{div}\overline{u}'_0. \tag{6.112}$$

From (6.110)–(6.112), using known results [74], [33], we obtain

$$|\sigma|\|\overline{u}'\|_{L_p} + \|\overline{u}'\|_{W_p^{(2)}} + \|\nabla P'\|_{L_p} \leq C(\|\overline{f}'\|_{L_p} + \|\overline{g}'\|_{W_p^{(1)}} + |\sigma|\sum_{k=1}^{3}\|g'_k\|_{L_p}). \tag{6.113}$$

Since $\overline{u}' = \overline{u}$ and $P' = P$ for $x \in \Omega - \Omega_h$, from (6.113) with consideration of (6.108) and (6.109) we deduce that

$$\begin{aligned}|\sigma|\|\overline{u}\|_{L_p(\Omega-\Omega_h)} &+ \|\overline{u}\|_{W_p^{(2)}(\Omega-\Omega_h)} + \|\nabla P\|_{L_p(\Omega-\Omega_h)} \\ &\leq C(\|\overline{f}\|_{L_p(\Omega)} + \|\overline{u}\|_{W_p^{(1)}(\Omega)} + \|P\|_{L_p(\Omega)} + |\sigma|\|\overline{B}\|_{L_p(\Omega)}).\end{aligned} \tag{6.114}$$

From (6.103) and (6.114) we obtain

$$\begin{aligned}|\sigma|\|\overline{u}\|_{L_p(\Omega)} &+ \|\overline{u}\|_{W_p^{(2)}(\Omega)} + \|\nabla P\|_{L_p(\Omega)} \\ &\leq C(\|\overline{f}\|_{L_p(\Omega)} + \|\overline{u}\|_{W_p^{(1)}(\Omega)} + \|P\|_{L_p(\Omega)} + |\sigma|\|\overline{B}\|_{L_p(\Omega)}).\end{aligned} \tag{6.115}$$

It remains to eliminate superfluous terms from the right side of (6.115). Suppose first that $p \geq 2$. It is then possible to use an energy estimate: taking the inner product of (6.1) with $\overline{u}^*$ and integrating over Ω, we obtain

$$\sigma\|\overline{u}\|_H^2 + \|\overline{u}\|_{H_1}^2 = \int_\Omega \overline{f}\cdot\overline{u}\,dx. \tag{6.116}$$

Separating the real and imaginary parts in (6.116), we find that

$$\operatorname{Re}\sigma\|\overline{u}\|_H^2 + \|\overline{u}\|_{H_1}^2 = \operatorname{Re}\int_\Omega \overline{f}\cdot\overline{u}^*\,dx \tag{6.117}$$

$$\operatorname{Im}\sigma\|\overline{u}\|_H^2 = \operatorname{Im}\int_\Omega \overline{f}\cdot\overline{u}^*\,dx. \tag{6.118}$$

If $\operatorname{Re}\sigma \geq 0$, applying the Cauchy-Schwarz-Bunyakovskiĭ inequality, from (6.117) and (6.118) we then obtain

$$\|\overline{u}\|_H \leq \frac{\|\overline{f}\|_H}{\operatorname{Re}\sigma}; \qquad \|\overline{u}\|_H \leq \frac{\|\overline{f}\|_H}{|\operatorname{Im}\sigma|}. \tag{6.119}$$

Since $|\sigma| \leq \sqrt{2}\max\{|\operatorname{Re}\sigma|, |\operatorname{Im}\sigma|\}$, from (6.119) we obtain

$$\|\overline{u}\|_H \leq C\|\overline{f}\|_H/|\sigma|. \tag{6.120}$$

If $\operatorname{Re}\sigma < 0$, then $|\operatorname{Im}\sigma| \geq |\operatorname{Re}\sigma|\tan\theta$, $|\operatorname{Im}\sigma| \geq \sin\theta|\sigma|$, and therefore from (6.118) we again obtain (6.120), which is thus valid for all $\sigma \in \Sigma_{0,\theta}$.

We now estimate the pressure P in L_2. Suppose $\pi(x)$ is a solution of the Neumann problem

$$\Delta\pi = P; \quad \left.\frac{\partial\pi}{\partial n}\right|_S = 0; \quad \int_\Omega \pi\,dx = 0. \tag{6.121}$$

Since the pressure is defined up to an additive constant, the latter may be assumed chosen so that

$$\int_\Omega P\,dx = 0. \tag{6.122}$$

Problem (1.21) is then uniquely solvable, and we have

$$\|\pi\|_{W_2^{(2)}} \leq C\|P\|_{L_2} \tag{6.123}$$

and the imbedding theorem following from it

$$\|\nabla\pi\|_{L_6} \leq C\|P\|_{L_2}; \qquad \|\nabla\pi\|_{L_4(S)} \leq C\|P\|_{L_2}. \tag{6.124}$$

Taking the inner product of (6.1) with $\Delta\pi^*$ and integrating over Ω, we obtain

$$\int_\Omega |P|^2\,dx = -\int_\Omega \overline{f}\cdot\nabla\pi^*\,dx + \int_S \operatorname{curl}\overline{u}\times\nabla\pi^*\cdot\overline{n}\,ds. \tag{6.125}$$

Estimating the right side of (6.125) with the help of the Hölder inequality and (6.124), we find that

$$\|P\|_{L_2(\Omega)} \leq C(\|\overline{f}\|_{L_{6/5}} + \|\operatorname{curl}\overline{u}\|_{L_{4/3}(S)}). \tag{6.126}$$

We now use the known inequalities

$$\|\operatorname{curl}\overline{u}\|_{L_{4/3}(S)} \le \varepsilon\|\overline{u}\|_{W_p^{(2)}} + C_\varepsilon\|\overline{u}\|_{L_2}; \qquad (p > \tfrac{6}{5}); \tag{6.127}$$

$$\|P\|_{L_p} \le \varepsilon\|P\|_{W_p^{(1)}} + C_\varepsilon\|P\|_{L_2}, \tag{6.128}$$

which are satisfied for any $\varepsilon > 0$. With their help from (6.126) we obtain (possibly with another constant C_ε)

$$\|P\|_{L_2(\Omega)} \le C\|\overline{f}\|_{L_p} + \varepsilon\|\overline{u}\|_{W_p^{(2)}} + C_\varepsilon\|\overline{u}\|_{L_2}. \tag{6.129}$$

Aside from (6.127) and (6.128), the following inequalities of the same type hold:

$$\|\overline{u}\|_{W_p^{(1)}} \le \varepsilon\|\overline{u}\|_{W_p^{(2)}} + C_\varepsilon\|\overline{u}\|_{L_2} \tag{6.130}$$

$$\|B\|_{L_p} \le \varepsilon\|\overline{u}\|_{L_p} + C_\varepsilon\|\overline{u}\|_{L_2}. \tag{6.131}$$

In writing out (6.131), we have used the circumstance that $\|\overline{u}\|_{L_p}$ by Lemma 6.1 is an equivalent norm in $W_p^{(1)}$ of the vector B. The validity of (6.4)–(6.6) for $p \ge 2$ now follows immediately from (6.115), (6.120), and (6.129)–(6.131), where ε is chosen sufficiently small. In the case $p < 2$ the required result is obtained essentially by passing to the adjoint problem. Namely, suppose $\overline{v}$ is the solution of problem (6.1)–(6.3) with right side $\overline{f}_0 = |\overline{u}|^{p-2}\overline{u} \in L_{p'}$. Since $p' > 2$, by applying the part of the theorem already proved

$$\|\overline{v}\|_{L_{p'}(\Omega)} \le \frac{C}{|\sigma|}\|\overline{f}_0\|_{L_{p'}} = \frac{C}{|\sigma|}\|\overline{u}\|_{L_p}^{p-1}. \tag{6.132}$$

We take the inner product of (6.1) with $\overline{v}^*$ and integrate over the domain. After integrating by parts and applying the Hölder inequality, we find that

$$\int_\Omega |\overline{u}|^p\,dx = \int_\Omega \overline{f}\cdot\overline{v}^*\,dx \le \|\overline{f}\|_{L_p}\cdot\|\overline{v}\|_{L_{p'}}. \tag{6.133}$$

Now (6.5) follows immediately from (6.132) and (6.133).

We now estimate the pressure in L_p. For this we introduce the function π as a solution of the Neumann problem

$$\Delta\pi = |P|^{p-2}P + K; \quad \left.\frac{\partial\pi}{\partial n}\right|_S = 0; \quad \int_\Omega \pi\,dx = 0, \tag{6.134}$$

where the constant K is defined so that problem (6.134) is solvable:

$$K = -\frac{1}{m_3\Omega}\int_\Omega |P|^{p-2}P\,dx. \tag{6.135}$$

The function π satisfies

$$\|\pi\|_{W_{p'}^{(2)}} \le C\|P\|_{L_p}^{p-1}. \tag{6.136}$$

We take the inner product of (6.1) with $\nabla\pi^*$ and integrate over Ω. Integrating by parts and considering (6.122), in place of (6.125) we obtain

$$\int_\Omega |P|^p\,dx = -\int_\Omega \overline{f}\cdot\nabla\pi^*\,dx + \int_S \operatorname{curl}\overline{u}\times\nabla\pi^*\cdot\overline{n}\,ds \tag{6.137}$$

By applying the Hölder inequality, (6.136), and the imbedding theorem for $p > \frac{3}{2}$, we conclude that

$$\begin{aligned}\|P\|_{L_p}^p &\le \|\overline{f}\|_{L_p}\cdot\|\nabla\pi\|_{L_{p'}} + \|\operatorname{curl} u\|_{L_{\frac{2}{3}p}(S)}\cdot\|\nabla\pi\|_{L_{2p'/(3-p')}}(S)\\ &\le C(\|\overline{f}\|_{L_p} + \|\operatorname{curl}\overline{u}\|_{L_{\frac{2}{3}p}(S)})\|P\|_{L_p}^{p-1}.\end{aligned} \tag{6.138}$$

We thus have

$$\|P\|_{L_p} \le C(\|\overline{f}\|_{L_p} + \|\operatorname{curl}\overline{u}\|_{L_{\frac{2}{3}p}(S)}). \tag{6.139}$$

After this (6.4)–(6.6) are derived from (6.115) in exactly the same way as was done above in the case $p \ge 2$, but in place of (6.120) we use (6.5) and in place of (6.127)

$$\|\operatorname{curl}\overline{u}\|_{L_{\frac{2}{3}p}(S)} \le \varepsilon\|\overline{u}\|_{W_p^{(2)}} + C_\varepsilon\|\overline{u}\|_{L_p}, \tag{6.140}$$

while in place of (6.129)–(6.131) we use analogous inequalities with $\|\overline{u}\|_{L_2}$ replaced by $\|\overline{u}\|_{L_p}$.

In the case $p = \frac{3}{2}$ the proof is similar; in place of $2p'/(3-p)$ it is possible to take any number $q > 1$. Finally, if $p < \frac{3}{2}$, then in the foregoing argument it suffices to replace $L_{2p'/(3-p')}(S)$ by $C(S)$ and $L_{\frac{2}{3}p}(S)$ by $L_1(S)$.

The a priori estimates (6.4)–(6.6) have thus been proved. We shall not pause to consider the proof of existence of a solution, since it is carried out by well-known schemes: the existence of a generalized solution is first established (see §5), and it is then proved that estimates (6.4)–(6.6) hold for it; in this regard see [78], [1], [82], and [83].

THEOREM 6.2. *Suppose the conditions of Theorem* 6.1 *are satisfied and the right side f in equation* (6.1) *has the form*

$$f = \sum_{k=1}^{3}\frac{\partial f^k}{\partial x^k}. \tag{6.141}$$

Then the solution (u, P) *of boundary value problem* (6.1)–(6.3) *satisfies*

$$\|u\|_{W_p^{(1)}} \leq C \sum_{k=1}^{3} \|f^k\|_{L_p}; \tag{6.142}$$

$$\|u\|_{L_q} \leq \frac{C}{|\sigma|^{1-\beta}} \sum_{k=1}^{3} \|f^k\|_{L_p}; \quad 2\beta = \frac{n}{q} - \frac{n}{p} + 1; \quad 0 < \beta \leq \frac{1}{2}; \quad n = 2, 3 \tag{6.143}$$

$$\|P\|_{L_p} \leq C \sum_{k=1}^{3} \|f^k\|_{L_p}, \tag{6.144}$$

where the constant C *is the same for all* $\sigma \in \Sigma_{0,\alpha}$.

PROOF. It may be assumed that the f^k are smooth vector-valued functions which vanish near the boundary S. We begin with the more elementary estimate (6.143). The solution of (6.1)–(6.3) with right side (6.141) has the form

$$u = \sum_{k=1}^{3} R_\sigma D_k f^k; \qquad R_\sigma = (\sigma I - \Pi\Delta)^{-1}; \qquad D_k = \frac{\partial}{\partial x_k}. \tag{6.145}$$

Using Lemma 5.4, it is not hard to see that the operator $R_\sigma D_k \colon L_p \to S_q$ has the adjoint operator $-D_k R_\sigma^* \colon S_{q'} \to L_{p'}$, where q' and p' are the dual indices. Therefore, (6.143) follows from the coincidence of the norms of a linear operator and its adjoint, Lemma 5.7, and the multiplicative inequality (see (4.10))

$$\|u\|_{W_p^{(1)}} \leq c\|u\|_{L_{q'}}^{\beta} \cdot \|u\|_{W_{q'}^{(2)}}^{1-\beta}. \tag{6.146}$$

To derive (6.142) and (6.144) we use the method of hydrodynamic potentials developed in [75], [76], [79], and [43], in which the system (6.1)–(6.3) is considered for $\sigma = 0$, but the results carry over essentially without change to nonzero $\sigma \in \Sigma_{0,\alpha}$.

We shall assume that the vector-valued functions f^k are defined in the entire space and are equal to zero outside the domain Ω. A solution w, Q of system (6.1)–(6.3) vanishing at infinity can be represented in the form of volume potentials

$$w_i(x) = \int_{R^3} u_i^s(x,y) f_s(y)\,dy; \qquad Q(x) = \int_{R^3} q^s(x,y) f_s(y)\,dy; \tag{6.147}$$

$$u^s = \operatorname{curl}\operatorname{curl} V^s; \qquad V^s(x,y) = \frac{1}{\gamma}\varphi(z) l_s; \qquad z = \gamma|x-y|;$$

$$\varphi(z) = \frac{1}{4\pi z}(1-e^{-z}); \tag{6.148}$$

$$q^s(x,y) = (\Delta - \gamma^2)\operatorname{div} V^s = -\frac{1}{4\pi}\frac{\partial}{\partial x_S}\frac{1}{|x-y|}. \tag{6.149}$$

The complex number γ is the square root of σ, with $\operatorname{Re}\gamma > 0$ for $\sigma \in \Sigma_{0,\alpha}, \sigma \neq 0$; l_1, l_2, l_3 are the coordinate unit vectors. As σ varies in $\Sigma_{0,\alpha}$ the parameters γ and z run over a sector $0 \leq \arg\gamma \leq (\pi-\alpha)/2$, and in this sector the function φ is bounded together with all its derivatives; moreover, $\varphi^{(k)}(z) = O(z^{-k-1})$ as $|z| \to \infty$.

Substituting into (6.147)

$$f_s = \sum_{k=1}^{3} \frac{\partial f_s^k}{\partial x_k}, \tag{6.150}$$

taking the derivatives from f_s^k onto the kernels u_i^s and q^s, and applying the Calderón-Zygmund theorem on singular integrals, one can obtain

$$\|w\|_{W_p^{(1)}} + \|Q\|_{L_p} \leq C\sum_{k=1}^{3} \|f^k\|_{L_p}. \tag{6.151}$$

Here and below constants depending only on α (and possibly on S) are denoted by C. Estimate (6.151) also follows from Theorem 6.1, since the operators D_k commute with R_σ if $\Omega = R^3$.

Seeking a solution of system (6.1)–(6.3) in the form

$$u = w + v; \qquad P = Q + q \tag{6.152}$$

to determine v and q we obtain the problem

$$\sigma v - \Delta v = -\nabla q; \qquad \operatorname{div} v = 0; \qquad v/s = -w/s. \tag{6.153}$$

Following [37], we seek a solution in the form of double-layer potentials:

$$v_i(x) = \int_S K_{ij}(x,y)\varphi_j(y)\,dS_y; \qquad q(x) = \int_S K_j(x,y)\varphi_j(y)\,dS_y; \tag{6.154}$$

$$K_{ij}(x,y) = -\left[\delta_j^k q^i + \frac{\partial u_k^i}{\partial y_j} + \frac{\partial u_j^i}{\partial y_k}\right] n_k(y); \tag{6.155}$$

$$K_j(x,y) = 2\frac{\partial}{\partial x_k} q^j(x,y) n_k(y).$$

Here $n_k(y)$ is the kth coordinate of the unit outer normal at the point $y \in S$. The functions φ_j here must be found by solving the system of integral equations

$$\frac{1}{2}\varphi_i(\xi) + \int_S K_{ij}(\xi, \eta)\varphi_j(\eta)\, dS_\eta = -w_i(\xi); \qquad \xi \in S. \tag{6.156}$$

For the kernels K_{ij} we have

$$\begin{gathered} |K_{ij}(\xi, \eta)| \le \frac{C}{|\xi - \eta|}; \\ |K_{ij}(\xi, \eta) - K_{ij}(\xi', \eta)| \le \frac{C|\xi - \xi'|}{R^2}; \\ R = \min\{|\xi - \eta|, |\xi' - \eta|\}. \end{gathered} \tag{6.157}$$

This can easily be derived from (6.148), (6.149), and (6.155) by using the above-indicated properties of the function φ and the estimate

$$|(\xi_k - \eta_k)n_k(\eta)| \le C|\xi - \eta|^2, \tag{6.158}$$

following from the fact that the boundary S belongs to the class C^2.

From (6.151) by known imbedding theorems [59] we obtain

$$\begin{aligned} \|w\|^p_{W_p^{1/p'}} &\equiv \int_S |w|^p\, ds + \int_S \int_S \frac{|w(x') - w(x)|^p}{|x' - x|^{p+1}}\, dS_x\, dS_{x'} \\ &\le C\left(\sum_{k=1}^{3} \|f^k\|_{L_p}\right)^p. \end{aligned} \tag{6.159}$$

Further, as in [43], it can be established that the system (6.156) has a solution $\varphi_1 \in L_p(S)$ which is uniquely determined by the added condition

$$\int_S \varphi_k n_k\, dS = 0 \tag{6.160}$$

and is subject to the estimate

$$\|\varphi_i\|_{L_p(S)} \le C\|w\|_{L_p(S)}. \tag{6.161}$$

We now prove that $\varphi_i \in W_p^{1/p'}(S)$ and

$$\|\varphi_i\|_{W_p^{1/p'}(S)} \le C\sum_{k=1}^{3} \|f^k\|_{L_p}. \tag{6.162}$$

In view of (6.159), for this it suffices to establish that the integrals in (6.156) belong to the class $W_p^{1/p'}(S)$. We consider one of them, omitting the indices for brevity:

$$\psi(\xi) = \int_S K(\xi, \eta)\varphi(\eta)\, dS_\eta. \tag{6.163}$$

Applying the Hölder inequality and setting $\alpha = 1 - 3/4p$, we obtain

$$|\psi(\xi) - \psi(\xi')|^p \leq I^{p-1} \int_S |K(\xi, \eta) - K(\xi', \eta)|^{(1-\alpha)p} |\varphi(\eta)|^p \, dS_\eta;$$

$$I = \int_S |K(\xi, \eta) - K(\xi', \eta)|^{\alpha p'} \, dS_\eta. \tag{6.164}$$

For the integral I from (6.157) we immediately obtain

$$I \leq C|\xi - \xi'|^{2-\alpha p'}. \tag{6.165}$$

We now consider the integral

$$J = \int_S \int_S \frac{|K(\xi, \eta) - K(\xi', \eta)|^{(1-\alpha)p}}{|\xi - \xi'|^{-(2-\alpha_{p'})(p-1)+p+1}} \, dS_\zeta \, dR_{\zeta'}. \tag{6.167}$$

We prove that it converges. Indeed, using (6.157), we obtain

$$J = \int_S \int_S \frac{dS_\xi \, dS_{\xi'}}{R^{3/2}|\xi - \xi'|^{3/2}} < \infty. \tag{6.168}$$

Further, from (6.164) by integrating with respect to ξ and ξ' with consideration of (6.165) and (6.168), we get

$$\int_S \int_S \frac{|\psi(\xi') - \psi(\xi)|^p}{|\xi - \xi'|^{p+1}} \, dS_\xi \, dS_{\xi'} \leq C \int_S |\varphi(\eta)|^p \, dS_\eta. \tag{6.169}$$

Estimate (6.162) is obtained from (6.156), (6.159), (6.161), and (6.169).

We rewrite the expression (6.154) for the function q with consideration of (6.155) and (6.149) in the form

$$q(x) = \frac{\partial^2 g_{kj}}{\partial x_k \partial x_j}; \quad g_{kj}(x) = -\frac{1}{2\pi} \int_S \frac{n_k(y)\varphi_j(y)}{|x - y|} \, dSy. \tag{6.170}$$

The functions g_{kj} are harmonic in Ω. Using the familiar formula for the normal derivative of a single-layer potential, we find for $\xi \in S$

$$\frac{\partial g_{kj}(\xi)}{\partial n_\xi} = n_k(\xi)\varphi_j(\xi) - \frac{1}{2\pi} \int_S \frac{n_S(\xi)(\xi_S - \eta_S)}{|\xi - \eta|^3} n_k(\eta)\varphi_j(\eta) \, dS_\eta. \tag{6.171}$$

The integral in (6.171) can be estimated in exactly the same way as (6.163). It therefore follows from (6.171) that

$$\left\| \frac{\partial g_{kj}}{\partial n} \right\|_{W_p^{1/p'}(S)} \leq C \sum_{k=1}^{3} \|f^k\|_{L_p(\Omega)}. \tag{6.172}$$

Applying a known estimate of a harmonic function in $W_p^{(2)}$ (see [70] and [1]), we obtain

$$\|q\|_{(L_p)\Omega} \leq C_1 \sum_{k,j} \|g_{kj}\|_{W_p^{(2)}} \leq C_2 \sum_{k=1}^{3} \|f^k\|_{L_p}. \tag{6.173}$$

Comparing (6.151), (6.152), and (6.173), we obtain (6.144).

We now consider in Ω the boundary value problem

$$\sigma v - \Delta v = \frac{\partial g^k}{\partial x_k}; \quad v|_S = 0; \quad \sigma \in \Sigma_{0,\alpha}. \tag{6.174}$$

For the solution of it we have

$$\|v\|_{W_p^{(1)}} \leq C \sum_{k=1}^{3} \|g^k\|_{L_p}. \tag{6.175}$$

The proof does not differ from that given in [89] for the special case $\sigma = 0$, and we therefore do not present it here. Applying this estimate to equation (6.1) with right side (6.141) and using (6.144), we obtain (6.142), which completes the proof of Theorem 6.2.

THEOREM 6.3. *Suppose $S \in C^3$ and $f \in S_p^{(1)}$. Then, uniformly with respect to $\sigma \in \Sigma_{0,\alpha}$, the solution of problem* (6.1)–(6.3) *satisfies*

$$|\sigma| \|u\|_{S_p^{(1)}} \leq C\|f\|_{S_p^{(1)}}; \tag{6.176}$$

$$\|u\|_{S_p^{(3)}} \leq C\|f\|_{S_p^{(1)}}. \tag{6.177}$$

PROOF. This theorem is easily derived from Theorem 6.2. We write system (6.1)–(6.3) in the form

$$\sigma u - \Pi \Delta u = \Pi f = f. \tag{6.178}$$

We set $\Pi \Delta u = w$. From (6.178) it follows that $w/s = 0$. Therefore, applying the operator $\Pi \Delta$ to (6.178), we find that w is a generalized solution of the boundary value problem

$$\sigma w - \Delta w = \Delta f - \operatorname{grad} q; \quad \operatorname{div} w = 0; \quad w/_S = 0. \tag{6.179}$$

From this by Theorem 6.2 we obtain

$$\|w\|_{S_p^{(1)}} \leq C\|f\|_{S_p^{(1)}}. \tag{6.180}$$

We now obtain (6.177) by applying a known estimate in $W_p^{(3)}$ (see [82] and [43]) of the solution of the boundary value problem

$$\Delta u = w - \nabla q; \qquad \operatorname{div} u = 0; \qquad u/_S = 0. \tag{6.181}$$

The estimate (6.176) follows from (6.177), since $\sigma u = w + f$ according to (6.178). Theorem 6.3 is proved.

§7. Estimates of the leading derivatives of a solution of the linearized steady-state Navier-Stokes equations

It would be natural to expect that estimates of the derivatives of a solution of the steady-state problem follow from the estimates obtained in the preceding section for the resolvent of the linearized steady-state Navier-Stokes operator. In the case of L_2 this is actually the case (Theorem 3.3 is applicable). In the case of L_p a corresponding derivation is unknown; it might be that these estimates are not valid for all equations of the class in question. In the case of the Navier-Stokes equations all this makes it necessary to give a direct proof (see Theorems 7.1–7.3). As P. E. Sobolevskiĭ has indicated (see [73]), it is true that an estimate of the resolvent can be derived from a coerciveness inequality.

Thus, in a bounded domain Ω with boundary $S \in C^2$ we consider the linearized Navier-Stokes equations

$$\frac{\partial \overline{u}}{\partial t} - \Delta u = -\nabla P + \overline{f}; \tag{7.1}$$

$$\operatorname{div} \overline{u} = 0 \tag{7.2}$$

with boundary condition

$$u/_S = 0. \tag{7.3}$$

We shall henceforth consider either T-periodic solutions $u(x, t+T) \equiv u(x, t)$, $P(x, t+T) \equiv P(x, t)$, assuming that the vector of mass forces f is T-periodic, or the problem with the initial condition

$$u/_{t=0} = a. \tag{7.4}$$

We eliminate the ambiguity in the determination of the pressure by assuming that

$$\int_\Omega P\,dx = 0.$$

THEOREM 7.1. *Suppose $f(x,t)$ is a T-periodic function of class $L_p(Q_T)$ ($p > 1, Q_T = \Omega\times[0,T]$). Then problem* (7.1)–(7.3) *has a unique T-periodic solution possessing in Q_T generalized derivatives*(6) $\partial\overline{u}/\partial t$, $\partial^2\overline{u}/\partial x_i\partial x_k \in L_p(Q_T)$, *and*

$$\left\|\frac{\partial u}{\partial t}\right\|_{L_p(Q_T)} + \|D_x^2 u\|_{L_p(Q_T)} + \|\nabla P\|_{L_p(Q_T)} \leq C\|f\|_{L_p(Q_T)}. \tag{7.5}$$

The proof of this theorem is essentially a tracing of the proof of Theorem 6.1. The next assertion is an analogue of the basic Lemma 6.3.

(6)It is possibly not superfluous to emphasize that we consider Q_T as a torus by identifying the times 0 and T. Correspondingly, in defining generalized derivatives it is necessary to subject test functions to the condition of periodicity in t.

LEMMA 7.1. *Let $v(x,t)$ and $q(x,t)$ $(x=(\xi,z),\xi_1=x_1,\xi_2=x_2,z=x_3,\xi\in R^2,z\geq 0)$ constitute a solution T-periodic in time of the system*

$$\frac{\partial v}{\partial t}-\Delta v=-\nabla q+\overline{f}; \tag{7.6}$$

$$\operatorname{div} v=g=\sum_{k=1}^{3}\frac{\partial g_k}{\partial x_k}; \tag{7.7}$$

$$v/_{z=0}=0. \tag{7.8}$$

Assume that $v,q,\overline{f},g_k$, and g decay sufficiently fast as $|x|\to\infty$ (are compactly supported in x, for example). Then

$$\left\|\frac{\partial v}{\partial t}\right\|_{L_p(Q_T^+)}+\|D_x^2 v\|_{L_p(Q_T^+)}+\|\nabla q\|_{L_p(Q_T^+)}$$
$$\leq C\left(\|f\|_{L_p(Q_T^+)}+\|\nabla g\|_{L_p(Q_T^+)}+\sum_{k=1}^{3}\left\|\frac{\partial g_k}{\partial t}\right\|_{L_p(Q_T^+)}\right); \tag{7.9}$$

$$Q_T^+=R_3^+\times[0,T]; \qquad R_3^+=\{x\colon x\in R^3,x_3\geq 0\}.$$

PROOF. Let $\theta(t)$ be a compactly supported, infinitely differentiable function on the entire axis $-\infty<t<\infty$ satisfying the conditions $\theta(t)=1$ for $t\in[0,T]$, $\theta(t)=0$ for $t>2T$ or $t<-T$, and $|\theta(t)|\leq 1$. We set

$$u=\theta v; P=\theta q; \quad f_0=\theta f+\theta' u; \quad g_0=\theta g; \quad g_{k0}=\theta g_k. \tag{7.10}$$

From (7.6)–(7.8) we then obtain

$$\frac{\partial u}{\partial t}-\Delta u=-\nabla P+f_0 \tag{7.11}$$

$$\operatorname{div} u=g_0=\sum_{k=1}^{3}\frac{\partial g_{k0}}{\partial x_k} \tag{7.12}$$

$$u/_{z=0}=0. \tag{7.13}$$

As in Lemma 6.3, we represent the vector $\overline{u}(\xi,z,t)$ in the form

$$u=u_0+\operatorname{grad}\varphi+w; \tag{7.14}$$

$$\frac{\partial u_0}{\partial t}-\Delta u_0=f_0; \quad u_0|_{z=0}=0; \quad \overline{u}_0|_{|x|\to\infty}=0; \tag{7.15}$$

$$\Delta\varphi=g_0-\operatorname{div} u_0=\sum_{k=1}^{3}\frac{\partial}{\partial x_k}(g_{k0}-u_{0k}); \quad \left.\frac{\partial\varphi}{\partial z}\right|_{z=0}=0; \tag{7.16}$$

$$\varphi/_\infty=0; \quad w=\operatorname{curl} G; \quad G=(-\psi_{x_2},\psi_{x_1},0);$$

$$\frac{\partial\Delta\psi}{\partial t}-\Delta^2\psi=0; \quad \psi|_{z=0}=0; \quad \left.\frac{\partial\psi}{\partial z}\right|_{z=0}=\varphi(\xi,0,t). \tag{7.17}$$

Here u_0 and ψ (and hence also φ) are assumed bounded for $t \in (-\infty, +\infty)$. By separation of variables we obtain

$$u_0(x,t) = \int_{-\infty}^{+\infty} d\tau \int_{R^3} e^{i(\tau t+\alpha x)} \frac{\hat{f}(\alpha,\tau)}{\alpha^2 - i\tau}\, d\alpha; \tag{7.18}$$

$$\alpha = (\alpha_1, \alpha_2, \alpha_3), \qquad \alpha x = \sum_{k=1}^{3} \alpha_k x_k;$$

$$\hat{f}(\alpha,\tau) = \frac{1}{(2\pi)^4} \int_{-\infty}^{+\infty} dt \int_{R^3} e^{-i(\tau t+\alpha x)} f_0(x,t)\, dx.$$

Here the vectors f_0 and u_0 are assumed extended to the half-space $z < 0$ in an odd manner.

We differentiate (7.18) with respect to α_j and α_k. We observe that the multiplier

$$\lambda_{jk}(\alpha) = \frac{\alpha_j \alpha_k}{\alpha^2 - i\tau} \tag{7.19}$$

on a lattice of the form $\alpha_j = l_j \omega$ ($j = 1,2,3; l_1, l_2$, and l_3 are integers) satisfies the condition of Marcinkiewicz's theorem on multipliers of Fourier series and, moreover, uniformly with respect to $\omega \in (0,\infty)$. This makes it possible to assert that the corresponding operator is bounded in $L_p(R^3 \times (-\infty, +\infty))$. We thus have

$$\|D_x^2 u_0\|_{L_p(Q_\infty^+)} \leq C \|f_0\|_{L_p(Q_\infty^+)}. \tag{7.20}$$

Estimating the derivative $\partial u_0/\partial t$ in a similar way (or using the fact that by (7.15) it is equal to $f_0 + \Delta u_0$), we obtain

$$\left\| \frac{\partial u_0}{\partial t} \right\|_{L_p(Q_\infty^+)} \leq C \|f_0\|_{L_p(Q_\infty^+)}. \tag{7.21}$$

Applying Theorem 1.6 and considering (7.20) and (7.21), for the function φ we obtain

$$\|D_x^3 \varphi\|_{L_p}^p \leq C^p \int_{-\infty}^{+\infty} [\|g_0(\cdot,t)\|_{W_p^{(1)}(R_3^+)}^p + \|f_0(\cdot,t)\|_{L_p(R_3^+)}^p]\, dt; \tag{7.22}$$

$$\left\| \nabla \frac{\partial \varphi}{\partial t} \right\|_{L_p} \leq C \left[\sum_{k=1}^{3} \left\| \frac{\partial g_{k0}}{\partial t} \right\|_{L_p} + \|f_0\|_{L_p} \right]. \tag{7.23}$$

We now proceed to problem (7.17). A solution of it can be written in the form

$$\psi(\xi, z, t) = \int_{-\infty}^{+\infty} e^{i\tau t}\, d\tau \int_{R^2} e^{i(\xi,\eta)} \frac{e^{\lambda_1 z} - e^{\lambda_2 z}}{\lambda_1 - \lambda_2} \hat{\varphi}(\eta, 0, \tau)\, d\eta, \tag{7.24}$$

where $\hat{\varphi}$ is the Fourier transform of the function $\varphi(\xi, 0, \tau)$,

$$\hat{\varphi}(\eta, 0, \tau) = \frac{1}{8\pi^3} \int_{-\infty}^{+\infty} dt \int_{R^2} e^{-i(\zeta,\eta)-i\tau t} \varphi(\xi, 0, t)\, d\xi, \tag{7.25}$$

and λ_1 and λ_2 are given by

$$\lambda_1 = -|\eta| = -\sqrt{\eta_1^2 + \eta_2^2}; \qquad \lambda_2 = -\sqrt{|\eta|^2 + i\tau}. \tag{7.26}$$

The root in (7.26) is uniquely determined by the condition $\operatorname{Re}\lambda_2 \le 0$.

We introduce a new function $q(x, t)$ by the equality

$$\frac{\partial \varphi}{\partial t} - \Delta\varphi = q. \tag{7.27}$$

Considering the boundedness of φ for $t \in (-\infty, +\infty)$ and the boundary conditions (7.16), we find the following expression for φ in terms q:

$$\begin{aligned} \varphi(\xi, z, t) &= \int_{-\infty}^{+\infty} e^{i\tau t}\, d\tau \int_{R^2} e^{i(\xi,\eta)} \hat{\varphi}(\eta, z, \tau)\, d\eta; \\ \hat{\varphi}(\eta, z, \tau) &= \frac{1}{\lambda_2} \cosh \lambda_2 z \int_0^{\infty} e^{\lambda_2 s} \hat{q}(\eta, s, \tau)\, ds \\ &\quad + \frac{1}{\lambda_2} \int_0^z \sinh \lambda_2 (z - s) \hat{q}(\eta, s, \tau)\, ds; \\ \hat{q}(\eta, s, \tau) &= \frac{1}{8\pi^3} \int_{-\infty}^{+\infty} e^{-i\tau t}\, dt \int_{R^2} e^{-i(\eta,\xi)} q(\xi, s, t)\, d\xi. \end{aligned} \tag{7.28}$$

From (7.28) we deduce that

$$\hat{\varphi}(\eta, 0, \tau) = \frac{1}{\lambda_2} \int_0^{\infty} e^{\lambda_2 s} \hat{q}(\eta, s, \tau)\, ds. \tag{7.29}$$

Substituting (7.29) into (7.24), we obtain

$$\Delta\psi = \int_0^{\infty} ds \int_{-\infty}^{+\infty} d\tau \int_{R^2} e^{i\tau t + i(\xi,\eta)} \frac{i\tau e^{\lambda_2(z+s)}}{\lambda_2(\lambda_2 - \lambda_1)} \hat{q}(\eta, s, \tau)\, d\eta \tag{7.30}$$

From (7.30) and (7.17) we obtain the following representation for Δw_1:

$$\Delta w_1 = -\Delta\psi_{x_1 z} = \int_0^{\infty} ds \int_{-\infty}^{+\infty} d\tau \int_{R^2} e^{i\tau t + i(\xi,\eta)} \frac{i\tau e^{\lambda_2(z+s)}}{\lambda_2 - \lambda_1} \hat{q}_{\xi_1}(\eta, s, \tau)\, d\eta. \tag{7.31}$$

We consider the inner integral in (7.31) for fixed z and s as an operator on the function $q_{\xi_1}(\xi, z, t)$. To prove that it is bounded in $L_p(R_+^3)$ we consider the multiplier

$$\Phi(\eta, \tau) = \frac{i\tau e^{\lambda_2(z+s)}}{\lambda_2 - \lambda_1} = (\lambda_2 + \lambda_1) e^{\lambda_3(z+s)}; \qquad \eta \in R^2;\ \tau \in R^1. \tag{7.32}$$

According to the theorem of Marcinkiewicz (see [53]), it suffices to verify that

$$(z+s)|\Phi(\eta,\tau)| \le M; \tag{7.33}$$

$$(z+s)\int_{\Delta_1}\left(\left|\frac{\partial\Phi}{\partial\eta_1}\right| d\eta_1 + \left|\frac{\partial\Phi}{\partial\eta_2}\right| d\eta_2 + \left|\frac{\partial\Phi}{\partial\tau}\right| d\tau\right) \le M; \tag{7.34}$$

$$(z+s)\int_{\Delta_1}\int_{\Delta_2}\left(\left|\frac{\partial^2\Phi}{\partial\eta_1\partial\eta_2}\right| d\eta_1 d\eta_2 + \left|\frac{\partial^2\Phi}{\partial\eta_2\partial\tau}\right| d\eta_2 d\tau + \left|\frac{\partial^2\Phi}{\partial\eta_2\partial\tau}\right| d\eta_2 d\tau\right) \le M; \tag{7.35}$$

$$(z+s)\int_{\Delta_1}\int_{\Delta_2}\int_{\Delta_3}\left|\frac{\partial^3\Phi}{\partial\eta_1\partial\eta_2\partial\tau}\right| d\eta_1\, d\eta_2\, d\tau \le M, \tag{7.36}$$

where Δ_k is an arbitrary interval of the form $(2^{\alpha_k}, 2^{\alpha_k+1})$ or $[-2^{\alpha_k+1}, -2^{\alpha_k})$ with nonnegative integers α_k, while the constant M does not depend on $\eta, \tau, z, s, \alpha_k$, or α.

For this we observe that the multiplier Φ satisfies

$$(z+s)|D_\eta^k D_\tau^l \Phi| \le \frac{M_0}{|\eta|^k|\tau|^l}; \qquad k+l \le 3;\ 0 \le l \le 1 \tag{7.37}$$

with an absolute constant M_0. We estimate, for example, $\partial\Phi/\partial\tau$. We have

$$\frac{\partial\Phi}{\partial\tau} = \frac{i}{2\lambda_2} e^{\lambda_2(z+s)}[1 + (\lambda_2+\lambda_1)(z+s)]. \tag{7.38}$$

From (7.38), considering the inequalities

$$|\mathrm{Re}\,\lambda_2| = \sqrt{\frac{|\lambda_2|^2 + |\eta|^2}{2}} \ge \frac{1}{\sqrt{2}}|\lambda_2|, \qquad |\lambda_1| \le |\lambda_2|,$$

we deduce that

$$\left|\frac{\partial\Phi}{\partial\tau}\right| \cdot (z+s) \le \frac{1}{2|\lambda_2|^2} e^{-u}(u\sqrt{2} + 4u^2);$$
$$u = -\mathrm{Re}\,\lambda_2(z+s) \ge 0. \tag{7.39}$$

Taking the maximum of the right side of (7.39) over u and noting that $|\lambda_2|^2 \ge |\tau|$, we obtain

$$\left|\frac{\partial\Phi}{\partial\tau}\right|(z+s) \le \frac{M_0}{|\tau|}; \qquad M_0 = \frac{1}{2}\max_{u\ge 0} e^{-u}(u\sqrt{2} + 4u^2). \tag{7.40}$$

The remaining inequalities can be established in a similar but more cumbersome manner. From (7.37) the estimates (7.33)–(7.36) can easily be derived. For example, for the left side of (7.36), using (7.37) and going over to polar coordinates in the plane of η_1 and η_2, we obtain

$$M_0 \int_{\Delta_1}\int_{\Delta_2} \frac{d\eta_1\, d\eta_2}{|\eta|^2} \cdot \int_{\Delta_3} \frac{d\tau}{|\tau|} \le \frac{\pi}{2} M \cdot \ln 2 \int_a^b \frac{d\eta_1}{|\eta_1|} = \frac{\pi}{2} M_0 (\ln 2)^2; \tag{7.41}$$

$$b = 2a; \qquad a = 2^{\alpha_k} + 2^{\alpha_s}.$$

Using the generalized Minkowski inequality ("the norm of an integral does not exceed the integral of the norm"), from (7.31) we now conclude that

$$\|\Delta w_1\|_{L_p(R^3_+)} \le C \int_0^\infty \frac{\|q_{x_1}(\cdot, s, \cdot)\|_{L_p(R^3_+)}}{z+s}\, ds; \qquad R^3_+ = R^2 \times (0, \infty). \tag{7.42}$$

From (7.42), by the boundedness of the Hilbert integral transform in $L_p(0, \infty)$ (see Theorem 1.4), we obtain

$$\|\Delta w_1\|_{L_p(Q^+_\infty)} \le C \|q_{x_1}\|_{L_p(Q^+_\infty)} \tag{7.43}$$

The following inequalities can be obtained in exactly the same way:

$$\|\Delta w_2\|_{L_p(Q^+_\infty)} \le C \|q_{x_2}\|_{L_p(Q^+_\infty)};$$

$$\|\Delta w_3\|_{L_p(Q^+_\infty)} \le C \sum_{k=1}^{2} \|q_{x_k}\|_{L_p(Q^+_\infty)}. \tag{7.44}$$

From (7.43) and (7.44) we obtain

$$\|\Delta \overline{w}\|_{L_p(Q^+_\infty)} \le C \sum_{k=1}^{2} \|q_{x_k}\|_{L_p(Q^+_\infty)}. \tag{7.45}$$

Further, estimating the right side of (7.45) according to (7.27) and (7.22), (7.23), we obtain

$$\|\Delta w\|_{L_p(\Omega^+_\infty)} \le C \left(\sum_{k=1}^{3} \left\| \frac{\partial g_{k0}}{\partial t} \right\|_{L_p} + \|\nabla g_0\|_{L_p} + \|f_0\|_{L_p} \right). \tag{7.46}$$

The estimate (7.9) is now derived from (7.46) just as in Lemma 6.3 if we further note that for any function $h(\xi, z, t)$ T-periodic in t we have

$$\|h\|_{L_p(Q^+_T)} \le \|\theta h\|_{L_p(Q^+_\infty)} \le \sqrt{3} \|h\|_{L_p(Q_T)}. \tag{7.47}$$

Lemma 7.1 is proved.

PROOF OF THEOREM 7.1 The derivation of the a priori estimate (7.5) from Lemma 7.1 almost literally repeats the proof of Theorem 6.1, and we therefore omit it for fear of boring the reader.

The existence of a solution satisfying (7.5) can be proved by known methods (see [37] and [107]). We therefore here only sketch briefly one of the possible paths of the argument. In the case $p = 2$ the existence of a solution follows from Theorems 6.1 and 3.3. For $1 < p < 2$ we obtain the same result by approximating $f \in L_p$ by vector-valued functions in L_2 and using the estimate (7.5) on passing to the limit.

Suppose now that $p > 2$. We denote by S'_p the space of vector-valued functions T-periodic in t with values in S_p. On the set D'_p dense in S'_p of vector-valued functions vanishing on the boundary of Ω and having generalized derivatives u_t, $D_x^2 u \in L_p(Q_T)$, we define the operator L by setting

$$Lu = u_t - \Pi\Delta u. \tag{7.48}$$

The estimate (7.5) shows that L is a closed operator. On the linear manifold $D'_{p'}, p' = p/(p-1)$, we now define the operator L^* by setting

$$L^* v = -v_t - \Pi\Delta v. \tag{7.49}$$

The operator L^* reduces to L by the change $t \to -t$. Since $p' < 2$, L^{*-1} exists by what has been proved. The invertibility of L will be established if we show that its adjoint operator is L^*. Of course, the latter constitutes the principal technical difficulty here. By Lemma 5.4 it suffices to show that the equality

$$\int_0^T \int_\Omega Lu \cdot v \, dx \, dt = \int_0^T \int_\Omega u \cdot g \, dx \, dt$$

for all $u \in D'_p$ and a given $g \in S'_{p'}$ implies that $v \in D'_{p'}$ and $L^* v = \xi$. This assertion can be extracted from considerations of [107].

Using Theorem 7.1, it is not hard to estimate the leading derivatives of a solution of system (7.1)–(7.3) if the vector-valued function f is sufficiently smooth. Thus, by successively differentiating (7.1)–(7.3) with respect to time and then applying Theorem 7.1 and the known estimate in $W_p^{(l)}$ of a solution of the steady-state problem, we prove the following theorem.

THEOREM 7.2. *Suppose* $S \in C^{(m)}, m = 2, 4, \ldots$. *Then a* T*-periodic solution of problem* (7.1)–(7.3) *satisfies*

$$\sum_{k+2l \le m} [\|D_x^k D_t^l u\|_{L_p(Q_T)} + \|D_x^{k+1} D_t^{c-1} P\|_{L_p(Q_T)}]$$
$$\le C \sum_{k+2l \le m-2} \|D_x^{2k} D_t^l f\|_{L_p(Q_T)} \tag{7.50}$$

We obtain an estimate of a solution of the problem with initial data by applying Theorems 3.2 and 7.1.

THEOREM 7.3. *A solution of problem* (7.1)–(7.4) *satisfies*

$$\|u_t\|_{L_p(Q_T)} + \|D_x^2 u\|_{L_p(Q_T)} \leq C(\|f\|_{L_p(Q_T)} + |||a|||_p), \tag{7.51}$$

where $p > 1$, $T > 0$, and the constant C depends only on the domain Ω and on p;

$$|||a|||_p = \inf(\|v_t\|_{L_p(Q_T)} + \|D_x^2 v\|_{L_p}); \qquad v(x,0) = a(x). \tag{7.52}$$

The estimate (7.51) (with the space of initial data made concrete) was obtained by Solonnikov [78].

CHAPTER II

Stability of Fluid Motion

§1. Stability of the motion of infinite-dimensional systems

We consider the nonlinear equation

$$\frac{dv}{dt} = F(v, t) \tag{1.1}$$

in a Banach space X, and we assume that some solution $v_0(t)$ of it defined for $t \geq 0$ is known; we henceforth call v_0 the *basic solution.* If in (1.1) we make the change $v = v_0 + u$, then for the perturbation u we obtain

$$\frac{du}{dt} = F(v_0 + u, t) - F(v_0, t) \equiv f(u, t). \tag{1.2}$$

Equation (1.2) is called a *nonlinear perturbation equation.* We shall assume that the perturbation is given at the initial time:

$$u(0) = u_0. \tag{1.3}$$

We suppose (at least for a set of vectors u_0 dense in some neighborhood of zero of the space X) that the Cauchy problem (1.2), (1.3) is uniquely solvable (at least for small t: $0 < t < \tau(u_0)$). The following mapping is thus defined:

$$u(t) = U^t u_0. \tag{1.4}$$

In the important special case where the right side of (1.2) does not depend on t explicitly this mapping is a (nonlinear partial) semigroup

$$U^{t_1+t_2} u_0 = U^{t_1} U^{t_2} u_0; \qquad (0 \leq t_1, t_2;\ t_1 + t_2 \leq \tau(u_0) > 0). \tag{1.5}$$

Many definitions are known of stability of a solution (or, what is the same, stability of the zero solution of (1.2)) on the basis of the initial data. The essence of the matter, however, is that it is required, first of all, that the Cauchy problem (1.2), (1.3) for "small" u_0 have a solution defined for all $t > 0$ and, second, that the mapping U^t be continuous. Of course, there are many definitions of continuity, and to each there corresponds a concept of stability. In place of normed spaces it is possible to consider topological linear spaces; this makes it possible, in particular, to include in the present scheme the concept of "continuous dependence on initial data" and the η-stability defined below.

The stability of the basic solution of equation (1.1) is henceforth identified with stability of the fixed point $u_0 = 0$ of the one-parameter family of mappings U^t. For the following definitions the connection of the mapping U^t with the differential equation is of no consequence. For the time being this enables us not to be concerned with the sense in which the vector-valued function (1.4) satisfies (1.2); in particular, various generalized solutions are included in the consideration.

DEFINITION 1.1. Suppose U is some space of functions $u(t)$ defined for $t \in [0, \infty)$ with values in the space X. The fixed point $u_0 = 0$ of the one-parameter family of mappings (1.4) is called *stable* (X, U) if U^t defines a mapping continuous at u_0 of some neighborhood of zero of X into the space U.

DEFINITION 1.2. Let $C_X = C([0, \infty), X)$ be the space of vector-valued functions $u(t)$ with values in the Banach space X which are continuous on $[0, \infty)$ and have finite norm

$$\|u(\cdot)\|_{C_X} = \sup_{0 \le t < \infty} \|u(t)\|_X. \tag{1.6}$$

We denote by C_X^0 the subspace in C_X consisting of vector-valued functions $u(t)$ such that $\|u(t)\|_X \to 0$ as $t \to \infty$.

Stability (X, C_X) is called *Lyapunov stability*; stability (X, C_X^0) is called *asymptotic stability*.

It is sometimes useful to consider stability (X_0, C_{X_1}) or $(X_0, C_{X_1}^0)$, where X_0 and X_1 are Banach spaces having a common dense set with X. We call it *Lyapunov* (respectively, *asymptotic*) *stability from X_0 to X_1*.

DEFINITION 1.3. *Exponential stability in the space X* is stability (X, C_X, σ), where $C_{X,\sigma}$ is the space of continuous vector-valued functions with norm

$$\|u(\cdot)\|_{C_{X,\sigma}} = \sup_{t \ge 0}(e^{\sigma t}\|u(t)\|X); \quad \sigma > 0 \tag{1.7}$$

DEFINITION 1.4. Let L_{p,σ,X_1} $(p \ge 1, \sigma > 0)$ denote the space of vector-valued functions $u(t)$ with finite norm

$$\|u(\cdot)\|_{L_{p,\sigma}} = \left[\int_0^\infty (e^{\sigma t}\|u(t)\|_{X_1})^p \, dt\right]^{1/p}. \tag{1.8}$$

Stability (X_0, L_{p,σ,X_1}) is called *exponential stability in the mean* (of degree p from X_0 to X_1).

In the infinite-dimensional case (say, for partial differential equations) the concept of Lyapunov stability is considerably richer than in the finite-dimensional case. This is connected with the possible inequivalence of norms. In particular, it can turn out (and hence does turn out!) that a solution of an equation is stable with one choice of metric and unstable with another choice. We present some examples.

EXAMPLE 1. We consider the Cauchy problem for the partial differential equation of first order

$$\frac{\partial u}{\partial t} - x\frac{\partial u}{\partial x} = 0; \qquad u|_{t=0} = \varphi(x). \tag{1.9}$$

The solution has the form

$$(U^t\varphi)(x) = \varphi(e^t x). \tag{1.10}$$

The zero solution of (1.9) is Lyapunov stable in $C(-\infty, +\infty)$, while in $L_p(-\infty, +\infty)$ $(p \geq 1)$ it is exponentially stable:

$$\|u^t\varphi\|_C = \|\varphi\|_C; \qquad \|u^t\varphi\|_{L_p} = e^{-t/p}\|\varphi\|_{L_p}. \tag{1.11}$$

Moreover, in the space C^1 there is no stability:

$$\begin{aligned}\|U^t\varphi\|_{C^1} &= \|U^t\varphi\|_C + \|D_x U^t\varphi\|_C \\ &= \|\varphi\|_C + e^t\|\varphi'\|_C \to \infty, \qquad (t \to \infty)\end{aligned} \tag{1.12}$$

if $\varphi \neq \text{const}$.

EXAMPLE 2. We consider the problem with initial data in the strip $|y| \leq 1$ on the (x, y) plane

$$\frac{\partial \Delta\psi}{\partial t} + y\frac{\partial \Delta\psi}{\partial x} = 0; \quad \psi\|_{t=0} = \varphi(x, y); \quad \psi|_{y=\mp 1} = 0. \tag{1.13}$$

This problem is obtained as a result of linearization about Couette flow (with a linear velocity profile and flow function $\psi_0 = y^2/2$) of the two-dimensional equations of motion of an ideal fluid in a flat pipe. Writing $\Delta\psi = -\omega$ and $\Delta\varphi = -\omega_0$, we find for the vorticity the expression

$$\omega(x, y, t) = \omega_0(x - yt; y). \tag{1.14}$$

For the vorticity there is again Lyapunov stability in the space C,

$$\max_{x,y} |\omega(x, y, t)| = \max_{x,y} |\omega_0(x, y)|, \tag{1.15}$$

while in C^1 there is no stability:

$$\begin{aligned}\max_{x,y} |\omega_y(x, y, t)| &= \max_{x,y} |-t\omega_{0x}(x - yt, y) + \omega_0(x - yt, y)| \\ &\geq t\max_{x,y} |\omega_{0x}(x, y)| - \max_{x,y} |\omega_0(x, y)| \to \infty \qquad (t \to \infty),\end{aligned} \tag{1.16}$$

if ω_0 depends on x.

An analogous result is obtained in the case of Poiseuille flow $v_r = v_\theta = 0$, $v_z = 1 - r^2$ in a circular pipe $0 \leq r \leq 1$ (r, θ, z are cylindrical coordinates; the quantity ω_θ / r plays the role of the vorticity) and also in the case of a flow in a ring $r_1 < r < r_2$ with flow function $\psi_0 = Ar^2 + \beta \ln r$.

It may be supposed that this phenomenon occurs for all flows of an ideal incompressible fluid and not only in a linear problem but also in a nonlinear problem; in the three-dimensional case, probably, by perturbing a steady-state solution of the Euler equations an arbitrarily small amount, it is possible to arrive at a time-dependent flow with a vorticity increasing as $t \to \infty$. (Is this not connected with the circumstance that the very effective condition of stability for two-dimensional flows derived by V. I. Arnol'd has so far not been verified for any three-dimensional flow?)

From what is presented below it follows that for flows of a viscous fluid and solutions of parabolic equations, on the other hand, in the stable case peturbations die out together with their derivatives of all orders. As we know, it is true that the derivatives of a solution of a parabolic equation for $t = 0$ may for arbitrarily smooth data have discontinuities on the boundary of the domain at the initial time (where the initial and boundary values meet); in order that there be smoothness it is necessary to subject the data to certain consistency conditions. Since we are interested in the behavior of solutions as $t \to \infty$, it is natural to ignore these discontinuities. Here it is convenient to introduce the following definition.

DEFINITION 1.5. The fixed point $u_0 = 0$ of the one-parameter family of mappings U^t (1.4) is called *η-stable* (X_0, X_1) if 1) for any $a \in X_0$ in some neighborhood of u_0: $\|a\|_{X_0} < \delta_0$, the vector $U^t a \in X_1$ for any $t > 0$; and 2) for any pair of numbers $\eta, \varepsilon > 0$ there exists $\delta > 0$ such that $\|a\|_{X_0} < \delta$ implies that $\|U^t a\|_{X_1} < \varepsilon$ for $t > \eta$. If in addition to this $\|U^t a\|_{X_1} \to 0$ as $t \to 0$, then we say that *asymptotic stability* (X_0, X_1) holds.

EXAMPLE. We consider the problem with initial data for the one-dimensional heat equation

$$u_t - u_{xx} = 0; \quad u|_{t=0} = a(x) = \sum_{k=1}^{\infty} a_k \sin kx; \quad u|_{x=0,\pi} = 0. \tag{1.17}$$

The solution has the form

$$u(t,x) = \sum_{k=1}^{\infty} e^{-k^2 t} a_k \sin kx. \tag{1.18}$$

Clearly the zero solution is asymptotically η-stable $(L_2(0,\pi),\ W_2^{(n)}(0,\pi))$ for any $n \geq 0$:

$$\|D_x^n u(t,\cdot)\|_{L_2(0,\pi)} = \sqrt{\frac{\pi}{2}\sum_{k=1}^{\infty} e^{-2k^2 t} k^{2n} a_k^2}$$

$$\leq \mu(t)\|u(0,\cdot)\|_{L_2(0,\pi)}; \tag{1.19}$$

$$\mu^2(t) = \begin{cases} \left(\dfrac{n}{2te}\right)^n & 0 < t \leq \dfrac{n}{2}, \\ e^{-2t} & t \geq \dfrac{n}{2}. \end{cases}$$

Lyapunov stability occurs only for $n = 0$.

§2. Conditions for stability

Some theorems on stability of solutions of differential equations in a Banach space are proved in this section. These theorems together with the estimates of Chapter I lead to conditions of asymptotic stability in L_p and in $W_p^{(l)}$ for equations of parabolic type and the Navier-Stokes equations.

We consider a nonlinear differential equation

$$\frac{du}{dt} + Au = Ku \tag{2.1}$$

in a Banach space X, under the following assumptions.

I. Suppose Y is a Banach space having a common dense set with X and the operator A generates an analytic semigroup in X and in Y, i.e.,

$$\|(\sigma I - A)^{-1}\|_{X\to X} \leq C/|\sigma|; \tag{2.2}$$

$$\|(\sigma I - A)^{-1}\|_{Y\to Y} \leq C/|\sigma|, \tag{2.3}$$

uniformly with respect to $\sigma \in \sum_{\sigma_0,\theta}$; the sector

$$\sum_{\sigma_0,\theta} = \{\sigma\colon \theta \leq \arg(\sigma - \sigma_0) \leq \pi - \theta\}; \qquad 0 < \theta < \pi/2.$$

II. The imbedding operator $J_X\colon Y \to X$ of Y into X has fractional degree α $(0 < \alpha < 1)$ relative to the operator A from the right in the sense that

$$\|e^{-tA}J_X\|_{Y\to X} = \|e^{-tA}\|_{Y\to X} \leq \frac{C}{t^\alpha} e^{-\sigma_0 t} \qquad (t > 0). \tag{2.4}$$

III. The nonlinear operator K is analytic and homogeneous, i.e., it has the form

$$Ku = K_0(u_1, u_2, \dots, u_m)|_{u_1=u_2=\cdots=u_m=u}, \tag{2.5}$$

where the operator K_0 is linear in each of its arguments, and $m \geq 2$. We characterize the "differential properties" of the operator K_0 by requiring that

$$\|K_0(u_1, u_2, \ldots u_m)\|_Y \leq \|B_1 u_1\|_{Y_1} \cdots \|B_m u_m\|_{Y_m}, \tag{2.6}$$

where B_j ($j = 1, \ldots, m$) is a linear operator acting from X to the Banach space Y_j and having fractional degree β_j ($0 < \beta_j < 1$) relative to the operator A:

$$\|B_j e^{-tA}\|_{X \to Y_j} \leq \frac{C}{t^{\beta_j}} e^{-\sigma_0 t}. \tag{2.7}$$

Suppose further that the following condition is satisfied:

$$\sum_{j=1}^{m} \beta_j < 1 - \alpha. \tag{2.8}$$

We remark immediately that what follows carries over without difficulty to the case where the right side of (2.1) contains a sum (or a convergent series) of operators satisfying the condition just formulated; the condition of additivity could be replaced by a requirement of sufficient smoothness and it is also not hard to consider an operator K depending on time.

THEOREM 2.1. *Suppose conditions* I–III *are satisfied and the spectrum of the operator A lies within the right half-plane*: $\operatorname{Re} \sigma(A) > \alpha_0 > 0$. *Then the zero solution of equation* (2.1) *is Lyapunov stable and is, moreover, exponentially stable in the space X. Moreover, it is exponentially stable in the mean of degree r_0 from X to Y_0 if Y_0 is a Banach space having a common dense set with X, and relative to A the imbedding operator $B_0\colon X \to Y_0$ has fractional degree β_0: $0 < \beta_0 < 1 - \alpha$; r_0 is any number such that* $1 \leq r_0 < 1/\beta_0$.

PROOF. We introduce the Banach space Z consisting of continuous vector-valued functions $u(t)$ with values in X and having finite norm

$$\|u\|_Z = \sup_{t \geq 0} e^{\sigma_0 t} \|u(t)\|_X + \sum_{k=0}^{m} |u|_{r_k, \sigma_0}, \tag{2.9}$$

where the seminorm $|u|_{r_k, \sigma}$ is defined by

$$|u|_{r_k, \sigma} = \left\{ \int_0^\infty [e^{\sigma\tau} \|B_k u(\tau)\|_{Y_k}]^r \, d\tau \right\}^{1/r}. \tag{2.10}$$

The numbers r_k are subject to the condition $1 \leq r_k < 1/\beta_k$. We here choose r_k ($k = 1, \ldots, m$) so close to $1/\beta_k$ that

$$\frac{1}{p} = \sum_{k=1}^{m} \frac{1}{r_k} < 1 - \alpha. \tag{2.11}$$

We shall now prove that the zero solution of (2.1) is stable (X, Z)—this assertion is obviously equivalent to Theorem 2.1.

By inverting the left side of (2.1), we reduce the Cauchy problem for it to the integral equation

$$u(t) = u_0(t) + \int_0^t e^{-(t-\tau)A} Ku(\tau)\, d\tau \equiv (Nu)(t); \tag{2.12}$$

$$u_0(t) = e^{-tA}a; u(0) = a. \tag{2.13}$$

We first note that $u_0 \in Z$. Indeed, using condition (2.7), we get

$$\|u_0(t)\|_X \le Ce^{-\sigma_0 t}\|a\|_X; \tag{2.14}$$

$$\|B_k u_0(t)\|_{Y_k} \le \frac{C}{t^\beta k} e^{-\sigma_1 t}\|a\|_X, \tag{2.15}$$

where σ_1 is any number satisfying $\operatorname{Re}\sigma(A) > \sigma_1 > \sigma_0$. From (2.14) and (2.15) we obtain

$$\|u_0\|_Z \le C\|a\|_X. \tag{2.16}$$

We now show that the operator N defined in (2.12) is a contraction operator in some ball S_R of the space Z (of radius R with center at zero) if $\|a\|_X$ is sufficiently small. To this end we make a number of estimates.

We use the following representation of the semigroup e^{-tA}:

$$e^{-tA} = \frac{1}{2\pi i}\cdot\frac{1}{t}\int_\gamma e^{-\lambda t}R_\lambda^2\, d\lambda; \qquad R_\lambda = (\lambda I - A)^{-1}, \tag{2.17}$$

where γ is the boundary of the sector $\sum_{\sigma_0,\theta}$. Using this representation and setting $\lambda = \sigma_0 + re^{i\theta}$, we obtain

$$\|B_j e^{-tA} J_X\|_{Y\to Y_j} \le \frac{c}{t}e^{-\sigma_0 t}\int_0^\infty e^{-rt\cos\theta}\|B_j R_\lambda\|_{X\to Y_j} \cdot \|R_\lambda J_X\|_{Y\to X}\, dr. \tag{2.18}$$

From conditions (2.4) and (2.7) we obtain

$$\|B_j R_\lambda\|_{X\to Y_j} \le \frac{C}{|\lambda|^{1-\beta_j}}; \qquad \|R_\lambda J_X\|_{Y\to X} \le \frac{C}{|\lambda|^{1-\alpha}}. \tag{2.19}$$

With the help of (2.19), from (2.18) we deduce that

$$\|B_j e^{-tA} J_X\|_{Y\to Y_j} \le \frac{Ce^{-\sigma_0 t}}{t^\alpha j}; \qquad \alpha_j = \beta_j + \alpha. \tag{2.20}$$

We remark that $0 < \alpha_j < 1$ as a consequence of (2.8).

We introduce the operator $M\colon Z \to Z$ by setting

$$(Mu)(t) = \int_0^t e^{-(t-\tau)A}Ku(\tau)\, d\tau. \tag{2.21}$$

Using (2.20), we obtain

$$\|(B_k Mu)(t)\|_{Y_k} \leq \int_0^t \frac{Ce^{-\sigma_1(t-\tau)}}{(t-\tau)^{\alpha_k}} \|Ku(\tau)\|_Y \, d\tau. \tag{2.22}$$

Suppose first that the number p defined by (2.11) satisfies

$$1-(1-\alpha_k)p > 0. \tag{2.23}$$

Applying the theorem on potentials, from (2.22) we then obtain

$$|Mu|_{\bar{r}_k,\sigma_1} \leq C\left\{\int_0^\infty [e^{\sigma_1\tau}\|Ku(\tau)\|_Y]^p \, d\tau\right\}^{1/p}. \tag{2.24}$$

$$\bar{r}_k = \frac{p}{1-(1-\alpha_k)p}.$$

Applying the Hölder inequality and noting that $\bar{r}_k > r_k$ and $\sigma_1 > \sigma_0$, we deduce that

$$|Mu|_{r_k,\sigma_0} \leq C_0 |Mu|_{\bar{r}_k,\sigma_1}; \tag{2.25}$$

$$C_0 = [(\sigma_1-\sigma_0)r_k]^{(1/\bar{r}_k - 1/r_k)}.$$

Using (2.6) to estimate the right side of (2.24), we further obtain

$$|Mu|_{\bar{r}_k,\sigma_1} \leq C\left\{\int_0^\infty [e^{\sigma_0\tau}\|B_1u\|_{Y_1}]^p \cdots [e^{\sigma_0\tau}\|B_mu\|_{Y_m}]^p \, d\tau\right\}^{1/p}. \tag{2.26}$$

Here we use the arbitrariness in the choice of σ_1 and assume that $\sigma_1 < m\sigma_0$. Estimating the right side of (2.26) with the help of the Hölder inequality with exponents $\lambda_1 = r_1/p, \ldots, \lambda_m = r_m/p$ and applying (2.25), we get

$$|Mu|_{r_k,\sigma_0} \leq C|u|_{r_1,\sigma_0} \cdots |u|_{r_m,\sigma_0} \leq C\|u\|_Z^m. \tag{2.27}$$

Inequality (2.27) is also valid in the case where condition (2.23) is not satisfied. Indeed, we denote the right side of (2.22) by $Ce^{-\sigma_1 t}\psi(t)$ and set $e^{\sigma_1 t}\|Ku(\tau)\|_Y = \varphi(\tau)$.

We have

$$\psi(t) = \int_0^t \frac{\varphi(\tau)}{(t-\tau)^{\alpha_k}} \, d\tau = \int_0^t \varphi^{1/2}(\tau) \cdot \frac{\varphi^{1/2}(t)}{(t-\tau)^{1/2\alpha_k}} \cdot \frac{dt}{(t-\tau)^{1/2\alpha_k}}. \tag{2.28}$$

Applying the Hölder inequality with exponents $\lambda_1 = 2p$, $\lambda_2 = 2$, and $\lambda_3 = 2p'$, and cancelling $\psi^{1/2}$, from (2.28) we obtain

$$\psi(t) \leq C_1\|\varphi\|_{L_p(0,\infty)} \cdot t^{1/p' - \alpha_k}; \qquad C_1 = (1-\alpha_k p')^{-1/p'}. \tag{2.29}$$

Since $1/p' - \alpha_k = [p(1-\alpha_k)-1]/p > 0$, from (2.29) we obtain (2.26) with any $\bar{r}_k$, and with it also (2.27).

Further, from (2.4) and (2.21) we derive the estimate

$$\|Mu(t)\|_X \leq \int_0^t \frac{Ce^{-\sigma_0(t-\tau)}}{(t-\tau)^\alpha}\|Ku(\tau)\|_Y\, d\tau. \tag{2.30}$$

Estimating the right side of (2.30) with the help of the Hölder inequality, we obtain

$$e^{\sigma_0 t}\|Mu(t)\|_X \leq \left\{\int_0^t [e^{m\sigma_0\tau}\|Ku(\tau)\|_Y]^p\, d\tau\right\}^{1/p} \cdot \left\{\int_0^t \frac{e^{-\sigma_0 p'\tau}}{(t-\tau)^{\alpha p'}}\, d\tau\right\}^{1/p'} \tag{2.31}$$

The second factor on the right side of (2.31) is a bounded function of t; this follows from the estimate

$$\int_0^t \frac{e^{-\sigma_0 p'\tau}\,d\tau}{(t-\tau)^{\alpha p'}} \leq \int_0^1 \max_{t\geq 0} \frac{t^{1-\alpha p'}e^{-\sigma_0 p' st}}{(1-s)^{\alpha p'}}\, d\tau = \left(\frac{1-\alpha p'}{\sigma_0 p' e}\right)^{1-\alpha p'} \int_0^1 (1-s)^{-\alpha p'} s^{\alpha p'-1}\, ds. \tag{2.32}$$

Convergence of the integrals is ensured by the condition $0 < \alpha p' < 1$, which follows from (2.11). Estimating the first factor on the right side of (2.31) in exactly the same way as in the derivation of (2.27), we get

$$e^{\sigma_0 t}\|Mu(t)\|_X \leq C\|u\|_Z^m. \tag{2.33}$$

From (2.27) and (2.33) we obtain

$$\|Mu\|_Z \leq C\|u\|_Z^m. \tag{2.34}$$

Suppose now that $u_1, u_2 \in Z$. We take up the estimate of $\|Nu_1 - Nu_2\|$. We have

$$\begin{aligned} Nu_1 - Nu_2 &= \int_0^t e^{-(t-\tau)A}[Ku_1(\tau) - Ku_2(\tau)]\, d\tau \\ &= \int_0^t e^{-(t-\tau)A}[K_0(u_1 - u_2, u_1, \dots, u_1) \\ &\qquad + K_0(u_2, u_1 - u_2, u_1, \dots, u_1) \\ &\qquad + \dots + K_0(u_2, u_2, \dots, u_2, u_1 - u_2)]\, d\tau. \end{aligned} \tag{2.35}$$

Repeating the arguments in the derivation of (2.34), we obtain

$$\|Nu_1 - Nu_2\|_Z \leq C\|u_1 - u_2\|_Z \cdot (\|u_1\|_Z^m + \|u_2\|_Z \cdot \|u_1\|_Z^{m-1} + \dots + \|u_2\|_Z^m). \tag{2.36}$$

If $u_1, u_2 \in S_R$, then from (2.36) we get

$$\|Nu_1 - Nu_2\|_Z \leq q\|u_1 - u_2\|_Z, \tag{2.37}$$
$$q = CmR^m.$$

From (2.34) and (2.37) it follows that N is a contraction in S_R if:

$$q = CmR^m < 1; \qquad \|u_0\|_Z \le R - CR^m. \tag{2.38}$$

If R is chosen so small that the first inequality (2.38) is satisfied while in the second the first part is positive, then it can be arranged that the latter is satisfied if we assume that $\|a\|_X$ is sufficiently small and take account of (2.16).

According to the principle of contraction mappings, equation (2.13) has a unique solution in S_R. Moreover, it is not hard to show that uniqueness holds in the entire space Z. Indeed, suppose u_1 and u_2 are solutions of (2.13) and $v(t) = u_1 - u_2$ is their difference. If it is supposed that $v(t) \not\equiv 0$, then there exist $t_0 \ge 0$ and $\delta > 0$ such that $v(t) \ne 0$ for $t \in (t_0, t_0 + \delta)$. Now $v(t)$ for $t \ge t_0$ satisfies the equation

$$v(t) = \int_{t_0}^{t} e^{-(t-\tau)A}[Ku_1(\tau) - Ku_2(\tau)]\,d\tau. \tag{2.39}$$

We introduce the notation

$$\|v\|_{Z_T} = \sup_{t_0 \le t \le T} \|v(t)\|_X + \sum_{k=0}^{m} |v|_{r_k,\sigma_0,T}; \tag{2.40}$$

$$|v|_{r_k,T} = \left\{ \int_0^T \|B_k v(\tau)\|_{Y_k}^{r_k'}\, d\tau \right\}^{1/r_k}.$$

The previous method lets us deduce from (2.39) that

$$\|v\|_{Z_T} \le C(T - t_0)^{\gamma} \cdot \|v\|_{Z_T}; \qquad \gamma = 1/p' - \alpha > 0. \tag{2.41}$$

From (2.41) it follows that $v(t) = 0$ for $t \in [t_0, T]$ if T is sufficiently close to t_0. This contradiction proves that $v(t) = 0$ for $t > 0$, and uniqueness is proved.

If conditions (2.38) are satisfied the following estimate of a solution of (2.1) follows from the principle of contraction mappings and (2.16):

$$\|u\|_Z \le \frac{1}{1-q}\|u_0\|_Z \le C\|a\|_X, \tag{2.42}$$

from which stability also follows. Theorem 2.1 is thus proved.

It is not known whether Theorem 2.1 is true if in the basic condition (2.8) equality is admitted (apparently not). This is the case, however, in one important case of operators acting in L_p spaces.

We shall consider equation (2.1) under the following assumptions which describe in abstract form a situation characteristic for the Navier-Stokes equations and nonlinear parabolic equations.

I. The operator A generates an analytic semigroup in $L_P(\Omega, \mu, E)$ for any $p > 1$.

II. The operator K is analytic and homogeneous of degree $m \geq 2$. This means that it has the form

$$Ku = DLu = DL_0(u_1, u_2, \dots u_m)|_{u_1=u_2=\dots u_m=u}, \tag{2.43}$$

where L_0 is linear in each of its arguments and D is a linear operator. We suppose that for any $q > 1$

$$\|L_0(u_1, u_2, \dots, u_m)\|_{L_q(\Omega,\mu,E)} \leq \|B_1 u_1\|_{L_q\lambda_1} \cdots \|B_m u_m\|_{L_{q\lambda_m}(\Omega,\mu,E)}, \tag{2.44}$$

where $\lambda_1, \dots, \lambda_m > 1$ are numbers such that $\sum_1^m 1/\lambda_k = 1$. We assume that the operators $B_k\colon L_P \to L_{q_k}$ $(k = 1, \dots, m)$ relative to A have fractional degrees

$$\beta_{k,p,q} = \gamma_k + \eta\left(\frac{1}{p} - \frac{1}{q_k}\right) \tag{2.45}$$

for any $q_k \geq p > 1$ such that $0 < \beta_{k,p,q} < 1$; the numbers γ_k and η do not depend on p or q.

III. Suppose the operator $D\colon L_p \to L_q$ $(q \geq p)$ relative to A has fractional degree

$$\theta_{p,q} = \kappa + \eta\left(\frac{1}{p} - \frac{1}{q}\right) \tag{2.46}$$

in the sense that

$$\|e^{-tA}D\|_{L_p\to L_q} \leq C/t^{\theta}_{p,q}. \tag{2.47}$$

The number $\kappa, 0 < \kappa < 1$, does not depend on q or p.

IV. We require that

$$0 < \frac{1}{p_0} = \frac{1}{(m-1)\eta}\left(1 - \kappa - \sum_{k=1}^m \gamma_k\right) < 1. \tag{2.48}$$

Suppose there exist numbers r_k, q_k, r, $q > 1$ satisfying

$$r_k \geq p_0; \quad q_k \geq p_0; \quad \frac{1}{r_k} + \frac{\eta}{q_k} = \frac{\eta}{p_0} + \gamma_k; \quad \sum_{k=1}^m \frac{1}{r_k} \geq \frac{1}{p_0}; \tag{2.49}$$

$$\frac{1}{r} = \sum_{k=1}^m \frac{1}{r_k}; \qquad \frac{1}{q} = \sum_{k=1}^m \frac{1}{q_k}. \tag{2.50}$$

We further introduce the Banach space Z of vector-valued functions with finite norm

$$\|u\|_Z = \max_{t\geq 0} e^{\sigma_0 t}\|u(t)\|_{L_{p_0}} + \sum_{k=1}^m \|B_k u\|_{L_{q_k,r_k}}, \tag{2.51}$$

where we have used the notation

$$\|v\|_{L_{q,r}} = \left\{\int_0^\infty [e^{\sigma_0 t}\|v(t)\|_{L_q}]^r\,dt\right\}^{1/r}, \tag{2.52}$$

and σ_0 is a positive number.

THEOREM 2.2. *Suppose conditions* I–IV *are satisfied and the spectrum of the operator A lies within the right half-plane:* $\operatorname{Re}\sigma(A) > \sigma_0 > 0$. *Then the zero solution of equation* (2.1) *is asymptotically stable in* $L_{p_0} = L_{p_0}(\Omega, \mu, E)$. *Moreover, stability* (L_{p_0}, Z) *holds.*

PROOF. We again consider the integral equation (2.12). We first show that $u_0 \in Z$ and

$$\|u_0\|_Z \le C\|a\|_{L_{p_0}}. \tag{2.53}$$

Inequality (2.14) holds as before, and in the present case takes the form

$$\|u_0(t)\|_{L_{p_0}} \le Ce^{-\sigma_0 t}\|a\|_{L_{p_0}}. \tag{2.54}$$

Using Lemma 2.5 from Chapter I and considering conditions (2.46) and (2.49), we obtain

$$\|u_0\|_{L_{q_k,r_k}} \le C\|a\|_{L_{p_0}}. \tag{2.55}$$

Now (2.53) follows immediately from (2.54) and (2.55).

We further consider the operators Q, M, and M_0 defined by

$$(Qf)(t) = \int_0^t e^{-(t-\tau)A}Df(\tau)\,d\tau; \qquad (Mu)(t) = (QLu)(t); \tag{2.56}$$

$$M_0(u_1, u_2, \dots, u_m)(t) = QL_0(u_1, u_2, \dots, u_m)(t),$$

where D and L are the operators defined in (2.43). We shall show that the operator Q takes $L_{q,r}$ into Z continuously. Indeed, using conditions (2.46) and (2.49) and Lemma 2.6 of Chapter I,(7) we obtain

$$\|(Qf)(t)\|_{L_{p_0}} \le Ce^{-\sigma_0 t}\|f\|_{L_{q,r}}. \tag{2.57}$$

Lemma 2.2 of Chapter I applied to the operator Q with consideration of conditions (2.46) and (2.49) leads to the inequality

$$\|Qf\|_{L_{q_k,r_k}} \le C\|f\|_{L_{q,r}}. \tag{2.58}$$

From (2.57) and (2.58) we obtain the required estimate

$$\|Qf\|_Z \le C\|f\|_{L_{q,r}}. \tag{2.59}$$

Setting $\lambda_k = q/q_k$, from (2.59) and (2.44) we now obtain

$$\|M_0(u_1, u_2, \dots, u_m)\|_Z \le C\|B_1u_1\|_{L_{q_1,r_1}} \cdots \|B_mu_m\|_{L_{q_m,r_m}}. \tag{2.60}$$

---

(7)This is obviously also true in the case where the semigroup is multiplied by the operator D on the right.

We further argue as in the proof of Theorem 2.1: using (2.53) and (2.60), we conclude that the operator N of (2.12) is a contraction in some ball of the space Z if $\|a\|_{L_{p_0}}$ is sufficiently small. Theorem 2.2 follows directly from this.

We emphasize that Theorem 2.2 follows directly. If $\mu\Omega < \infty$, then under the conditions of the theorem there is stability (L_{p_0}, Z) also for

$$p_0 > \frac{(m-1)\eta}{1-\kappa-\sum_{k=1}^{m}\gamma_k}.$$

This follows directly from Theorem 2.1. Here it is essential that the rather restrictive conditions $r_k \geq p_0$ and $r \leq p_0$ drop out (see (2.49)).

We now apply Theorems 2.1 and 2.2 to the Navier-Stokes and parabolic equations.

Suppose a viscous incompressible fluid fills a bounded three-dimensional domain Ω with boundary $S \in C^2$, while the velocity vector on S and the mass forces are given and do not depend on time. Under these conditions suppose the Navier-Stokes equations have a steady-state solution $(v_0(x), P_0(x))$. The nonlinear perturbation equation can be written in the form (§5 of Chapter I)

$$\frac{du}{dt} + Au = Ku, \tag{2.61}$$

where A and K are operators defined by

$$Au = \nu A_0 u + Ru; \quad A_0 u = -\Pi\Delta u; \quad Ru = \Pi[(v_0, \nabla)u + (u, \nabla)v_0];$$

$$Ku = K_0(u, u); \qquad K_0(u, v) = -\Pi(u, \nabla)v \tag{2.62}$$

for any solenoidal vectors $u, v \in W_p^{(2)}$ vanishing on the boundary S. Π is the projection in L_p (orthogonal in L_2) onto the subspace S_p—the closure of the set of smooth solenoidal vectors in Ω vanishing near the boundary S. For any $p > 1$ the operators $-A$ and $-A_0$ generate an analytic semigroup in S_p.

THEOREM 2.3. *Suppose the spectrum of the operator A is situated within the right half-plane*: $\operatorname{Re}\sigma(A) > \sigma_0 > 0$. *Then the steady flow v_0 is asymptotically Lyapunov stable in S_p for $p > 3$. Moreover, if $u(0) = a$ has sufficiently small norm in S_p $(p > 3)$, then the solution $u(t)$ of* (2.61) *satisfies*

$$e^{\sigma_0 t}\|u(t)\|_{S_p} + \|u\|_{S_{q_1, r_1, t}} + \|D_x u\|_{L_{q_2, r_2, t}} \leq C\|a\|_{S_p}, \tag{2.63}$$

where

$$\|u\|_{S_{q,r,t}} = \left\{\int_0^t [e^{\sigma_0\tau}\|u(\tau)\|_{S_q}]^r \, d\tau\right\}^{1/r}. \tag{2.64}$$

The constant C does not depend on t. The numbers p_1, r_1, p_2, and r_2 must satisfy the conditions([8])

$$q_1 \geq p; \quad q_2 \geq p; \quad \frac{1}{q} = \frac{1}{q_1} + \frac{1}{q_2} > \frac{1}{p};$$
$$\frac{1}{r_1} > \frac{3}{2}\left(\frac{1}{p} - \frac{1}{q_1}\right);$$
$$\frac{1}{r_2} < \frac{3}{2}\left(\frac{1}{p} - \frac{1}{q_2} + \frac{1}{3}\right); \qquad r_1, r_2 > 1 \tag{2.65}$$

PROOF. For the operator K_0 we have the estimate

$$\|K_0(u_1, u_2)\|_{S_q} \leq C\|u_1\|_{S_{q_1}} \cdot \|\nabla u_2\|_{L_{q_2}}; \tag{2.66}$$
$$q > 1; \qquad \frac{1}{q_1} + \frac{1}{q_2} = \frac{1}{q}.$$

To derive (2.66) it suffices to note that the operator Π is bounded in S_q (Lemma 5.1 of Chapter I) and use the Hölder inequality. We now apply Theorem 2.1, in which we set $X = S_p$, $Y = S_q$, $Y_1 = S_{q_1}$, and $Y_2 = L_{q_2}$; we denote by B_1 the imbedding operator of the space S_p in S_{q_1} and by B_2 the operator $\nabla: v \to (\partial v_i/\partial x_k)$. Applying Lemma 5.7 of Chapter I and the multiplicative inequality (4.10), we then conclude that the operators B_1 and B_2 have the fractional degrees relative to A

$$\beta_1 = \frac{1}{2}\left(\frac{3}{p} - \frac{3}{q_1}\right); \qquad \beta_2 = \frac{1}{2}\left(\frac{3}{p} - \frac{3}{q_2} + 1\right). \tag{2.67}$$

By (2.65) we have $0 < \beta_1, \beta_2 < 1$. Further, from this same lemma we conclude that the imbedding operator of the space Y in the space X relative to A from the right has fractional degree α,

$$\alpha = \frac{3}{2}\left(\frac{1}{q} - \frac{1}{p}\right). \tag{2.68}$$

Here $0 < \alpha < 1/2$, since $p > 3$, and from (2.65) it follows that $q \geq p/2$.

Condition (2.8) is also satisfied; this follows immediately from (2.67), since $p > 3$. It remains to appeal to Theorem 2.1, and Theorem 2.3 is proved.

THEOREM 2.4. *Suppose the conditions of Theorem 2.3 are satisfied. Then the steady flow v_0 is stable in S_p for $p \geq 3$, and if $\|u(0)\|$ is sufficiently small, then the solution $u(t)$ of the nonlinear perturbation equation* (2.61) *satisfies*

$$e^{\sigma_0 t}\|u(t)\|_{S_p} + \|u\|_{S_{p_1, p_1, t}} \leq C\|a\|_{S_p}, \tag{2.69}$$

where $p_1 = 5p/3$ and C is a constant not depending on t or a.

([8])All these conditions are satisfied, for example, if $p = 4$, $q_1 = q_2 = 6$, $r_1 < 8$, and $r_2 < 8/5$.

PROOF. Using the solenoidal property of the vector u, for the operator $K_0(u,v)$ we can obtain the representation

$$K_0(u,v) = -\Pi(u,\nabla)v = -\Pi\left\{\frac{\partial}{\partial x_k}(u_i v_k)l_k\right\}, \tag{2.70}$$

where the l_k are the coordinate unit vectors in R^3. We use Theorem 2.2. Equality (2.70) shows that the operator K_0 can be given the form (2.43) if we define the operators(9) $L_0\colon (S_p \times S_p) \to L_{p/2}(\Omega, m_3, R^3 \otimes R^3)$ and $D\colon L_p(\Omega, m_3, R^3 \otimes R^3) \to S_q$ for any $p, q > 1$ by the equalities

$$L_0(u,v) = (u_i v_k)_{1\le i,k\le 3}; \tag{2.71}$$

$$D(a_{ik}) = -\Pi\left\{\frac{\partial a_{ik}}{\partial x_k} l_k\right\}. \tag{2.72}$$

The following inequality is here satisfied:

$$\begin{aligned}\|L_0(u_1,u_2)\|_{L_q} &= \left[\int_\Omega |u_1|^q \cdot |u_2|^q\,dx\right]^{1/q}\\ &\le \|u_1\|_{S_{q_1}} \cdot \|u_2\|_{S_{q_2}},\end{aligned} \tag{2.73}$$

if $q_1, q_2 > 1$ and $1/q_1 + 1/q_2 = 1/q$. This coincides with (2.44) if we set $m = 2$ and denote by B_k the imbedding operator of the space S_p in S_{q_k}. Here, by Lemma 5.7 and the multiplicative inequality (4.10) of Chapter I relative to the operator A, the operator B_k has fractional degree

$$\beta_{k,p,q} = \frac{3}{2}\left(\frac{1}{p} - \frac{1}{q_k}\right); \quad k = 1,2; \quad q_k \ge p > 1; \quad \beta_{k,p,q} < 1. \tag{2.74}$$

Further, by Theorem 6.3 of Chapter I the operator D has relative to A fractional degree

$$\theta_{p,q} = \frac{1}{2} + \frac{3}{2}\left(\frac{1}{p} - \frac{1}{q}\right). \tag{2.75}$$

Thus, conditions II and III are satisfied, with $\gamma_1 = \gamma_2 = 0, \kappa = 1/2$, and $\eta = 3/2$. From (2.48) we obtain $p_0 = 3$. Finally, setting $q_k = r_k = p_1 = 5$, we see that condition IV is satisfied. Therefore, for $p = 3$ Theorem 2.4 follows from Theorem 2.2. For $p > 3$ it is an obvious corollary of Theorem 2.3. Theorem 2.4 is thus proved.

Using the estimates of the preceding chapter, one can also investigate stability in $S_p^{(1)}$ and $S_p^{(2)}$ and prove that under the conditions of Theorem

(9)As usual, $S_p \times S_p$ denotes the Cartesian product of the space S_p. $R^3 \otimes R^3$ is the tensor square of the space R^3. If a basis in R^3 is fixed, then an element of $R^3 \otimes R^3$ can be realized as a matrix $\tau = (\tau_{ik})_{1\le i,k\le 3}$. We define the norm by the equality $\|\tau\| = \sqrt{\sum_{i,k=1}^3 a_{ik}^2}$.

2.3 the basic solution of the Navier-Stokes equations is Lyapunov stable in $S_p^{(1)}$ for $p \geq 3/2$ and in $S_p^{(2)}$ for any $p > 1$.

Theorems 2.1 and 2.2 are also applicable to the investigation of stability of solutions of parabolic equations. As an example we consider the equation

$$\frac{\partial u}{\partial t} + A_{2m}u = f(x)u^{k_0}(D_x u)^{k_1} \cdots (D_x^{2m-1}u)^k 2m - 1. \tag{2.76}$$

We assume that A_{2m} is an elliptic differential operator of order $2m$ in a bounded domain $\Omega \subset R^n$ and the conditions of Chapter I, §4, are satisfied; $k_0, \dots, k_{2m-1}$ are natural numbers, and f is a bounded measurable function.

THEOREM 2.5. *Suppose the spectrum of the operator A_{2m} lies in the right half-plane. Let*

$$0 < \frac{1}{p_0} = \frac{1 - \frac{1}{2m}s_1}{\frac{n}{2m}(s_0 - 1)}; \quad s_0 = \sum_j k_j; \quad s_1 = \sum_j jk_j. \tag{2.77}$$

Then the zero solution of equation (2.76) *is asymptotically stable in $L_p(\Omega)$ for any $p > \max(p_0, 1)$.*

PROOF. We set $X = L_p(\Omega)$, $Y = L_q(\Omega)$, and $Y_S = L_{P_s}(\Omega)$ (s runs over those integral values between 0 and $2m - 1$ for which $k_s > 0$), and we shall show that for an appropriate choice of q and p_s the conditions of Theorem 2.1 are satisfied. From Lemma 4.1 of Chapter I it follows that the operator $D_x^s \colon L_p \to L_{p_s}$ has relative to A_{2m} fractional degree

$$\beta_s = \frac{1}{2m}\left(\frac{n}{p} - \frac{n}{p_s} + s\right), \tag{2.78}$$

if $p_s \geq p$ and $\beta_s < 1$. The imbedding operator $J_x \colon X \to Y$ has relative to A fractional degree

$$\alpha = \frac{n}{2m}\left(\frac{1}{q} - \frac{1}{p}\right), \tag{2.79}$$

if $0 \leq \alpha < 1$.

An estimate of a nonlinear term of the type (2.6) can be obtained by means of the Hölder inequality with exponents $\lambda_s > 1$; we here introduce the additional condition $p_s \leq \lambda_s k_s q$. From the condition $p > p_0$ and (2.77) we then obtain

$$\sum_s k_s \beta_s < 1 - \alpha. \tag{2.80}$$

Thus, all the conditions of Theorem 2.1 are satisfied if it is possible to choose numbers $q, \lambda_s > 1$, and $p_s \geq p$ so that $0 \leq \alpha < 1$, $\beta_s < 1$, and

$p \le p_s \le \lambda_s k_s q$. The first of these conditions can be written in the form $1/p \le 1/q < 1/p + 2m/n$. By (2.80) the second condition is a consequence of the remaining conditions, while the last condition is satisfied if $s_0 q \ge p$. Both conditions on q reduce to the single condition

$$1/p \le 1/q < s_0/p, \tag{2.81}$$

since $(s_0-1)/p < (s_0-1)/p_0 = (2m/n)(1-s_1/2m) \le 2m/n$. Thus, suitable choices are any $q > 1$ satisfying (2.81) and any collection λ_s, p_s such that

$$\sum_s \frac{1}{\lambda_s} = 1; \qquad p \le p_s \le \lambda_s k_s q.$$

The theorem is thus proved.

In a number of cases Theorem 2.2 makes it possible to prove stability also in L_{p_0} (for example, for $m = 1$ and $k_0 = k_1 = 1$ for $n = 2$, and also for $n = 3$ if $f = \text{const}$).

Condition (2.77) is very restrictive, but in all probability it is necessary. If it is not satisfied, then one must study stability in $W^{(l)}_{p,A}$; for any equation of the form (2.76) one can find suitable p and $l \le 2m - 1$. We write out the corresponding limit exponent p_l:

$$1/p_l = 1/p_0 + l/n.$$

§3. Conditions for instability. Conditional stability

We shall consider equation (2.1), assuming that conditions I–III of Theorem 2.1 are satisfied. We further impose the following condition.

IV. The spectrum of the operator A can be represented as the union of closed sets σ_+ and σ_- with $\operatorname{Re}\sigma_+ \ge 0$ and $\operatorname{Re}\sigma_- < 0$.

A stronger condition is satisfied in applications to the Navier-Stokes and parabolic equations in bounded domains: the spectrum of the operator A is discrete—consists of a countable number of eigenvalues of finite multiplicity with the sole limit point at infinity.

Here the case where the operator has a connected spectral set containing points of both the left and right half-planes is left aside. We remark that also in this case instability can easily be proved by using the method of [38].

THEOREM 3.1. *Suppose conditions* I–III *of Theorem* 2.1 *and condition* IV *of the present section are satisfied. Suppose the set* σ_+ *is nonempty. Then the zero solution of equation* (2.10) *is unstable in* X.

This theorem follows from the next more general assertion. We denote by X_+ and X_- the subspaces in X invariant relative to the operator A

corresponding to the spectral sets σ_+ and σ_-. Let A_+ and A_- be the restrictions of A to X_+ and X_-, and let P_+ and P_- be the corresponding projections.

THEOREM 3.2. *Suppose the conditions of Theorem* 3.1 *are satisfied. Then there exists a manifold Y_- defined in some neighborhood of zero of the space X which is tangent to the subspace X_- at the point* 0 *and possesses the following properties.*

1. *For any point $u_0 \in Y_-$ there exists a solution $u(t)$, $u(0) = u_0$, of the Cauchy problem for equation* (2.1) *defined for all $t < 0$.*
2. *The manifold Y_- is invariant relative to translation along the trajectories of equation* (2.1): *if $u(0) \in Y_-$, then also $u(t) \in Y_-$ for all $t < 0$.*
3. $\|u(t)\|_X \to 0$ *exponentially as $t \to -\infty$.*

PROOF. We introduce the Banach space Z_1 consisting of vector-valued functions $u(t)$ continuous in t with values in X defined for all $t \leq 0$ and having the finite norm

$$\|u\|_{Z_1} = \sup_{t \equiv 0} e^{-\sigma_0 t} \|u(t)\|_X + \sum_{k=1}^{m} \|u\|_{r_k, \sigma_0}; \tag{3.1}$$

$$\|u\|_{r_k, \sigma_0} = \left\{ \int_{-\infty}^{0} [e^{-\sigma_0 t} \|B_k u(t)\|_{Y_k}]^{r_k} \, dt \right\}^{1/r_k}.$$

We choose the number σ_0 so that $\operatorname{Re} \sigma_- < -\sigma_0 < 0$, and we subject r_k to the inequalities

$$1 \leq r_k \leq \frac{1}{\beta_0}; \qquad \frac{1}{p} = \sum_{k=1}^{m} \frac{1}{r_k} < 1 - \alpha. \tag{3.2}$$

For any $a \in X$ we set $a_{\mp} = P_{\mp} a$; then $a = a_+ + a_-$. We consider the integral equation

$$u(t) = e^{-tA_-} a_- + \int_0^t e^{-(t-\tau)A_-} K u(\tau) \, d\tau + \int_{-\infty}^{t} e^{-(t-\tau)A_+} K u(\tau) \, d\tau \equiv (N_1 u)(t). \tag{3.3}$$

We shall show that for any a with sufficiently small norm this equation has a unique solution in Z_1. For this we first check that the operator N_1 acts continuously in Z_1. We consider the operator

$$(M_1 u)(t) = \int_{-\infty}^{t} e^{-(t-\tau)A_+} K u(\tau) \, d\tau. \tag{3.4}$$

We use the estimate

$$\|B_k e^{-sA} + J_X\|_{Y\to Y_k} \le Ce^{\eta s}/s^{\alpha}k; \qquad \alpha_k = \beta_k + \alpha, \tag{3.5}$$

where $s > 0$ and η is any positive number. It can be derived just as (2.20) was from the integral representation of the semigroup (2.17) in which the contour γ is chosen so that it lies entirely in the half-plane $\operatorname{Re}\lambda > -\eta$; we here assume that $\eta < \sigma_0$.

With the help of (3.5) we obtain

$$\|B_k(M_1 u)(t)\|_{Y_k} \le C\int_{-\infty}^{t} \frac{e^{\eta(t-\tau)}}{(t-\tau)^{\alpha_k}}\|Ku(\tau)\|_Y\,dt. \tag{3.6}$$

From (3.6) we derive the inequality

$$e^{-\sigma_0 t}\|B_k(M_1 u)(t)\|_{Y_k} \le C\int_{-\infty}^{t} e^{-m\sigma_0\tau}\|Ku(\tau)\|_Y \frac{d\tau}{(t-\tau)^{\alpha}k}. \tag{3.7}$$

Arguing as in the derivation of (2.34) and (2.36) from (2.20), we now get

$$\begin{aligned} &\|M_1 u\|_{Z_1} \le C\|u\|_{Z_1}^m;\\ &\|M_1 u_1 - M_1 u_2\|_{Z_1}\\ &\quad \le C\|u_1 - u_2\|_{Z_1}\\ &\quad\quad \times (\|u_1\|_{Z_1}^{m-1} + \|u_1\|_{Z_1}^{m-2}\cdot\|u_2\|_{Z_1} + \cdots + \|u_2\|_{Z_1}^{m-1}). \end{aligned} \tag{3.8}$$

The first two terms in (3.3) can be investigated just as (2.12) and (2.13) in the proof of Theorem 2.1. Arguing further as in the proof of Theorem 2.1, we conclude that for sufficiently small a_- the operator N_1 realizes a contraction mapping of some neighborhood of zero of the space Z_1 into itself, whence follows (local) unique solvability of (3.3).

It is also possible to arrive at this conclusion by applying the implicit function theorem.([10]) Indeed, we rewrite (3.3) in the form

$$F(a_-, u) \equiv u - N_1(a_-, u). \tag{3.9}$$

The operator F takes a neighborhood of zero of the space $X_- \times Z_1$ into Z_1 and is continuously differentiable (even analytic). Further, $F(0,0) = 0$ and $F_u(0,0) = I$, and hence the conditions of the implicit function theorem are satisfied. Therefore, (3.9) has a unique solution (in a neighborhood of

([10])D. V. Anosov pointed out to us the possibility of simplifying the proof of smoothness of the invariant manifold by the use of this theorem.

zero) $u = R(a_-)$, $R(0) = 0$, and the operator R is defined and continuously differentiable (even analytic) in some neighborhood of zero in X_-, and maps it into Z_1.

Any solution of (3.3) satisfies (at least in a generalized sense) equation (2.1). Indeed, from (3.3) it follows directly that for any $t_0 \leq t < 0$ we have

$$u(t) = e^{-(t-t_0)A}u(t_0) + \int_{t_0}^{t} e^{-(t-\tau)A}Ku(\tau)\,d\tau. \tag{3.10}$$

Setting $t = 0$ in (3.3), we obtain

$$a_+ = \int_{-\infty}^{0} e^{\tau A_+}Ku(\tau)\,d\tau \equiv \varphi(a_-). \tag{3.11}$$

Clearly φ is a continuously differentiable (even analytic) mapping of some neighborhood of zero of the space X_- into X_+, with $\varphi(0) = 0$ and $\varphi'(0) = 0$. Equation (3.11) defines the manifold Y_-. It is the diffeomorphic image of a neighborhood of zero in X_- and is tangent to this subspace at zero.

Applying the projection P_+ to (3.3), we obtain

$$u_+(t) = \int_{-\infty}^{t} e^{-(t-\tau)A_+}Ku(\tau)\,d\tau = \varphi(u_-(t)). \tag{3.12}$$

Hence, the manifold Y_- is invariant relative to translation along trajectories of equation (2.1). Theorem 3.2 is thus proved.

The following assertion is proved in a similar manner.

THEOREM 3.3. *Suppose conditions* I–III *of Theorem* 2.2 *are satisfied as well as condition* IV *of the present section. Then the zero solution of equation* (2.1) *is unstable in* L_{p_0}. *Moreover, there exists a subspace* Y_- *in* L_{p_0} *possessing the properties indicated in Theorem* 3.2.

Application of these theorems to the Navier-Stokes equations leads to the following result.

THEOREM 3.4. *Suppose the operator* A *in* (2.61) *has at least one point of the spectrum in the left half-plane. Then the steady flow* v_0 *is unstable in* S_p $(p \geq 3)$.

Moreover, it is not hard to prove that it cannot even be η-stable (C^k, S_p) for $p \geq 3$.

The theorems proved in this section admit a number of refinements. Under the conditions of Theorems 3.1–3.3 it is possible to show that all trajectories starting on a certain conical set issue from a neighborhood of zero; in the case where the operator A has no points of the spectrum on the imaginary axis the behavior of trajectories in a neighborhood of the steady-state regime can be described in considerable detail (in particular, in this case there exists an invariant manifold Y_+ which is stable as $t \to +\infty$). We shall not consider this here; such an investigation is carried out (by another method) in the more general case of a neighborhood of a periodic motion in the next chapter.

CHAPTER III

Stability of Periodic Motions

In this chapter the legitimacy of linearization in the problem of stability of periodic fluid motions is proved. A new method of proof for the case of steady motions is obtained at the same time. Forced oscillations and self-oscillations are considered separately.

§1. Formulation of the problem

We suppose that the data (the mass forces and the velocity on the boundary of the domain Ω occupied by the fluid) are such that the Navier-Stokes equations have a solution([11]) periodic in time with period T.

We denote the corresponding velocity vector by $v_0(t)$ and assume that it is sufficiently smooth (the smoothness conditions are given more precisely below). The change of variables $v = v_0 + u$ leads to the nonlinear perturbation equation

$$\frac{du}{dt} + A_0 u + B(t)u = Ku, \tag{1.1}$$

where we have used the notation (see Chapter I, §5)

$$\begin{aligned} &A_0 u = -\Pi\Delta u; \quad Ku = -K_0(u,u); \quad K_0(u,v) = \Pi(u,\nabla)v; \\ &K_0^0(u,v) = K_0(u,v) + K_0(v,u); \qquad B(t)u = K_0^0(u, v_0(t)). \end{aligned} \tag{1.2}$$

We further suppose that Ω is a bounded domain and its boundary $S \in C^2$. For simplicity it is assumed here that the coefficient of viscosity $\nu = 1$; this can be arranged by the change $u \to \nu u$, $v_0 \to \nu v_0$, $\nu t \to t$.

The method of linearization in the problem of stability of a periodic solution consists in the following. We consider the linearized pertubation

([11])General conditions for the existence of a periodic regime are given in [84], where a global existence theorem is formulated for generalized solutions in the three-dimensional and two-dimensional cases and also for smooth solutions in the two-dimensional case; generalized solutions in the two-dimensional case were also considered in [61] and [66].

equation

$$\frac{du}{dt} + A_0u + B(t)u = 0. \tag{1.3}$$

We shall seek solutions of (4.3) having the form

$$u(t) = e^{\sigma t}w(t) \tag{1.4}$$

with a vector-valued function w T-periodic in t. The complex values of the parameter σ for which (1.3) has nonzero solutions of the form (1.4) can be found by solving the spectral problem

$$\frac{dw}{dt} + A_0w + B(t)w + \sigma w = 0; \qquad w(t+T) \equiv w(t) \tag{1.5}$$

The collection of all such σ is called the *stability spectrum* of the basic flow v_0.

If the stability spectrum is situated within the left half-plane, then the basic flow is (asymptotically) stable. If even one point of the stability spectrum lies in the right half-plane, then the basic flow is unstable. If the right half-plane contains no points of the stability spectrum but there are some on the imaginary axis (the critical case), then it is not possible to draw a conclusion regarding stability by restricting attention to the linear approximation.

Generally speaking, the critical case is rarely encountered. Thus, if the basic flow depends on a parameter, then usually it occurs only for a discrete set of values of this parameter; the critical situation can be destroyed by an arbitrarily small perturbation of the basic flow. However, in investigating stability of self-oscillations it is always necessary to deal with the critical case. Below, this situation is treated separately (§6), and an infinite-dimensional analogue of the Andronov-Vitt theorem is proved.

The justification of the method of linearization is contained in the theorems of this chapter, which encompass a broad class of parabolic systems including the Navier-Stokes system.

To make the exposition specific and self-contained we speak below of stability in the energy space H_1. As in the case of steady flows, the same results hold also in the case of S_p ($p \geq 3$ in the three-dimensional case and $p \geq 2$ in the two-dimensional case) and in $S_p^{(1)}$ ($p \geq 3/2$ in the three-dimensional case and $p > 1$ in the two-dimensional case); it is also not hard to consider η-stability from S_p to $C^{k,\lambda}$

§2. The problem with initial data

We introduce the set M^T of vector-valued functions $u(t)$ of time $t \in [0, T]$ with values in $H = S_2$ such that $u(t) \in D(A_0)$ for all $t \in [0, T]$ and the vector-valued functions $u(t)$ and $(A_0u)(t)$ have strong derivatives of all orders with respect to t in H. We define the Hilbert space H_2^T as the closure of M^T in the metric generated by the inner product

$$(u,v)_{H_2^T} = \int_0^T \left[\left(\frac{du}{dt}, \frac{dv}{dt} \right)_H + (A_0u, A_0V)_H \right] dt + (u(T), v(T))_{H_1}. \quad (2.1)$$

Elements of H_2^T are vector-valued functions $u(t)$ such that for almost all $t \in [0,T]$ the strong derivative du/dt exists in H and $u(t) \in D(A_0)$.

LEMMA 2.1. *A vector-valued function $u \in H_2^T$ is strongly continuous in H_1 for $t \in [0,T]$ and*

$$\|u\|_{C_T(H_1)} \equiv \max_{0\leq t\leq T} \|u(t)\|_{H_1} \leq \|u\|_{H_2^T}. \quad (2.2)$$

PROOF. We use the identity

$$\frac{1}{2}(\|u(s)\|_{H_1}^2 - \|u(t)\|_{H_1}^2) = \int_t^s \left(\frac{du}{dt}, A_0u \right) d\tau, \quad (2.3)$$
$$(0 \leq t \leq s \leq T),$$

which is obvious for $u \in M^T$ and is obtained for any $u \in H_2^T$ by passing to the limit. From (2.3) we obtain

$$\|u(t)\|_{H_1}^2 \leq \int_t^s \left(\left\| \frac{du}{dt} \right\|_H^2 + \|A_0u\|_H^2 \right) d\tau + \|u(s)\|_{H_1}^2 \quad (2.4)$$

Setting $s = T$ here, we conclude that $u(t) \in H_1$ for all $t \in [0,T]$ and (2.2) holds. From (2.4) it follows that $\|u(t)\|_{H_1}$ is a continuous function of time for $t \in [0,T]$.

Further, suppose $\Phi \in D(A_0)$. Then for $u \in H_2^T$ we have

$$(u(s), \Phi)_{H_1} - (u(t), \Phi)_{H_1} = \int_t^s \left(\frac{du}{dt}, A_0\Phi \right)_H d\tau \quad (2.5)$$

with the help of the Cauchy-Schwarz-Bunyakovskiĭ inequality we deduce that

$$|(u(t), \Phi)_{H_1} - (u(s), \Phi)_{H_1}| \leq \|u\|_{H_2^T} \|A_0\Phi\|_H \sqrt{t-s}. \quad (2.6)$$

From (2.6) it follows that for $s \to t$

$$(u(s), \Phi)_{H_1} \to (u(t), \Phi)_{H_1}. \quad (2.7)$$

Since $D(A_0)$ is dense in H_1, from (2.2) and (2.7) by the Banach-Steinhaus theorem we conclude that $u(t)$ is strongly continuous in H_1. The lemma is proved.

LEMMA 2.2.

$$\|u(t)\|_{S_6} \leq C\|u\|_{H_2^T}; \qquad (0 \leq t \leq T); \tag{2.8}$$

$$\int_0^T \|u\|_{L_p}^{p_1}\,dt \leq C\|u\|_{H_2^T}^{p_1}; \qquad \int_0^T \|D_x u\|_{L_q}^{q_1}\,dt \leq C\|u\|_{H_2^T}^{q_1}; \tag{2.9}$$
$$p_1 = 4p/(p-6); \qquad q_1 = 4q/(3q-6),$$

where C depends only on the domain Ω, $6 < p < \infty$, and $2 < q \leq 6$.

PROOF. Inequality (2.8) follows from the Sobolev imbedding theorem and (2.2). According to Theorem 6.1 of Chapter I (see also [43] and [83]), we have

$$\|u\|_{W^{(2)}} \leq C\|A_0 u\|_H \qquad (u \in D(A_0)). \tag{2.10}$$

Using the Sobolev imbedding theorem, we obtain

$$\|u\|_{W_6^{(1)}} \leq C\|A_0 u\|_H. \tag{2.11}$$

Suppose now that $u \in H_2^T$. By the Hölder inequality

$$\|D_x u\|_{L_p} \leq \|D_x u\|_{L_2}^{3/q-1/2}\|D_x u\|_{L_6}^{3/2-3/q}; \qquad (2 \leq q \leq 6). \tag{2.12}$$

Raising (2.12) to the power $4q/(3q-6)$, integrating with respect to t from 0 to T, and using (2.2) and (2.11), we get the second inequality of (2.9).

To derive the first inequality of (2.9) we use the inequality

$$\|u(t)\|_{L_p}^{p_1} \leq C\|\nabla u(t)\|_{L_q}^{q_1}, \qquad q = 3p/(p+3), \tag{2.13}$$

which follows from the imbedding theorem, since the right side in (2.13) for vectors u vanishing on the boundary of the domain Ω defines a norm equivalent to the norm in $W_q^{(1)}$. Integrating (2.13) with respect to t and applying the second inequality of (2.9), we get the required relation. Lemma 2.2 is proved.

We now proceed to a consideration of the Cauchy problem for the differential equations (1.1) and (1.3) with the initial condition

$$u(0) = a; \qquad a \in H_1. \tag{2.14}$$

By a (generalized) solution of this problem on the time interval $[0, T]$ we mean a vector-valued function $u \in H_2^T$ satisfying the differential equation for almost all $t \in [0, T]$, and satisfying the initial condition (2.14) in the sense that

$$\|u(t) - a\|_{H_1} \to 0; \qquad (t \to 0). \tag{2.15}$$

We introduce the space $\widetilde{H}_2^T$—the closure of the set of smooth solenoidal vector fields $v(x,t)$ $(x \in \Omega, t \in [0,T])$ in the norm

$$\|v\|^2_{\widetilde{H}_2^T} = \int_0^T \left[\left\| \frac{\partial v}{\partial t} \right\|^2_{L_2(\Omega)} + \|v\|^2_{W_2^{(2)}(\Omega)} \right] dt + \left[\int_0^T \|v\|^4_{C(\Omega)} \, dt \right]^{1/2} + \left[\int_0^T \|v\|^4_{W_3^{(1)}(\Omega)} \, dt \right]^{1/2}. \tag{2.16}$$

If desired the last two terms can be omitted by estimating them in terms of the first term.

LEMMA 2.3. *Let $v_0 \in \widetilde{H}_2^T$. Then for any $a \in H_1$ there exists a unique generalized solution of the Cauchy problem* (1.3), (2.14).

PROOF. We first consider the Cauchy problem

$$\frac{du}{dt} + A_0 u = f(t); \qquad u(0) = a. \tag{2.17}$$

For a solution $u \in H_2^T$ of it we have

$$\begin{aligned} \|u\|^2_{H_2^t} \equiv \|u(t)\|^2_{H_1} + \int_0^t \left(\left\| \frac{du}{d\tau} \right\|^2_H + \|A_0 u\|^2_H \right) d\tau \\ = \|a\|^2_{H_1} + \int_0^t \|f(\tau)\|^2_H \, d\tau. \end{aligned} \tag{2.18}$$

Uniqueness of a generalized solution follows immediately from (2.18). With the help of this relation it is also not difficult to establish existence in the case $a \in H_1$. $\|f(t)\|_H \in L_2(0,T)$. For this it suffices, for example, to approximate f by smooth vector-valued functions or by finite segments of their Fourier expansion in the eigenvectors of the operator A_0. In these cases solvability of problem (2.17) is obvious, while (2.18) makes it possible to pass to the limit.

The solution of (2.17) can be written in the form

$$u(t) = e^{-tA_0} a + \int_0^t e^{-(t-\tau)A_2} f(\tau) \, d\tau. \tag{2.19}$$

It is now possible to reduce problem (2.3), (2.14) to the equivalent integral equation

$$\begin{gathered} u(t) = u_0(t) + (L_0 u)(t) \equiv (Lu)(t); \\ u_0(t) = e^{-tA_0} a; \qquad (L_0 u)(t) = -\int_0^t e^{-(t-\tau)A_0} B(\tau) u(\tau) \, d\tau. \end{gathered} \tag{2.20}$$

We construct a solution of (2.20) by successive approximations, taking u_0 as the initial approximation and setting

$$u_{n+1} = Lu_n. \tag{2.21}$$

We shall show that the sequence u_n converges in the metric of H_2^T. According to (2.18), $u_0 \in H_2^T$, and

$$\|u_0\|_{H_2^T} = \|a\|_{H_1}. \tag{2.22}$$

We now show that the vector-valued function $v_n = L^n u_0 \in H_2^T$ and

$$\|v_n(t)\|_{H_1}^2 \le \|v_n\|_{H_2^t}^2 \le \frac{m^n}{\sqrt{n!}} t^{n/2} \|a\|_{H_1}^2; \qquad (n = 0, 1, \dots). \tag{2.23}$$

Here the constant m does not depend on n and can be taken equal to

$$m = \left[\int_0^T q^2(\tau)\,d\tau\right]^{1/2}; \qquad q(\tau) = \max_{x\in\Omega} |v_0(\tau)|^2 + C_1^2 \|\operatorname{curl} v_0\|_{L_3}^2, \tag{2.24}$$

where C_1 is the norm of the imbedding operator of the space H_1 in L_6. Since $v_0 \in \widetilde{H}_2^T$, the quantity m is finite.

We use induction. Suppose (2.23) has already been proved for $0 \le n \le k$ $(k \ge 1)$. It is then also valid for $n = k + 1$. Indeed, from (1.2) we find that for any $u \in C_T(H_1)$

$$\|B(\tau)u(\tau)\|_H^2 \le \|v_0 \times \operatorname{curl} u + u \times \operatorname{curl} v_0\|_{L_2}^2 \le q(\tau)\|u\|_{H_1}^2, \tag{2.25}$$

where q is the function defined by (2.24).

Further, using (2.18) and (2.25), we deduce that

$$\|v_{k+1}(t)\|_{H_1}^2 \le \|v_{k+1}\|_{H_2^t}^2 \le \int_0^t q(\tau)\|v_k(\tau)\|_{H_1}^2\,d\tau. \tag{2.26}$$

If we now estimate the integral in (2.26) by means of (2.23) for $n = k$ and the Cauchy-Schwarz-Bunyakovskiĭ inequality, we arrive at (2.23) for $n = k + 1$. For $n = 0$ inequality (2.23) coincides with equality (2.22), already proved.

The estimate (2.23) is thus proved. It implies the absolute convergence in H_2^T of the series

$$\sum_{k=0}^{\infty} v_k = \sum_{k=0}^{\infty} L_0^k u_0 = \lim_{n\to\infty} u_n = u. \tag{2.27}$$

The vector-valued function u defined in (2.27) satisfies (2.20) and is hence a generalized solution of the Cauchy problem (1.3), (2.14). Lemma 2.3 is proved.

LEMMA 2.4. *The Cauchy problem for the nonlinear equation* (1.1) *with initial condition* (2.14) *has a unique generalized solution* $u \in H_2^T$ *for sufficiently small* a, *i.e., under the condition*

$$\|a\|_{H_1} \leq r, \tag{2.28}$$

where r *is a number depending only*([12]) *on* Ω *and* v_0. *If condition* (2.28) *is satisfied, then the solution of the Cauchy problem* (1.1), (2.14) *satisfies*

$$\|u\|_{C_T(H_1)} \leq \|u\|_{H_2^T} \leq C\|a\|_{H_1}. \tag{2.29}$$

PROOF. As in the proof of the preceding lemma, we reduce the problem to an integral equation.

$$u(t) = u_0(t) + (L_0 u)(t) + (Mu)(t), \tag{2.30}$$

where u and L_0 are the same as in (2.20), while the operator M is defined by

$$Mu = M_0(u, u);$$
$$M_0(u, v) = -\int_0^t e^{-(t-\tau)A_0} K_0(u(\tau), \nu(\tau))\, d\tau, \tag{2.31}$$

Using (1.2) and (2.18), for the operator M_0 we obtain, for any u, $v \in H_2^T$,

$$\begin{aligned}\|M_0(u, v)\|_{H_2^t}^2 &= \int_0^t \|K_0(u(\tau), v(\tau))\|^2\, d\tau \\ &\leq \int_0^t \int_\Omega |u|^2 |\nabla v|^2\, dx\, d\tau.\end{aligned} \tag{2.32}$$

Here

$$|\nabla v|^2 = \sum_{i,k=1}^{3} \left(\frac{\partial v_i}{\partial x_k}\right)^2.$$

With the help of the Hölder inequality and the imbedding theorem we deduce from (2.32) that

$$\|M_0(u, v)\|_{H_2^t}^2 \leq C_1 \int_0^t \|\nabla v\|_{L_3(\Omega)}^2\, d\tau \cdot \|u\|_{C^t(H_1)}^2, \tag{2.33}$$

where C_1 is the norm of the imbedding operator of H_1 in L_6. If we now apply Lemma 2.2, then from (2.33) we obtain

$$\|M_0(u, v)\|_{H_2^t} \leq C\|u\|_{H_2^t} \|v\|_{H_2^t}, \tag{2.34}$$

([12])If the coefficient of viscosity ν stands in front of the operator A_0 in (1.1) and it is assumed that ν_0 does not depend on ν, then $r \to 0$ as $\nu \to 0$.

where C depends only on the domain Ω. Further, for any u, $v \in H_2^T$ we have

$$(Mu)(t) - (Mv)(t) = M_0(u, u - v) + M_0(u - v, v). \tag{2.35}$$

From (2.34) and (2.35) we get

$$\|Mu - Mv\|_{H_2^t} \leq C(\|u\|_{H_2^t} + \|v\|_{H_2^t})\|u - v\|_{H_2^t}; \tag{2.36}$$

$$|Mu\|_{H_2^t} \leq C\|u\|_{H_2^t}^2. \tag{2.37}$$

Using Lemma (2.3), we reduce (2.30) to the equivalent operator equation in H_2^T

$$u = \varphi + (I - L_0)^{-1}Mu \equiv Qu; \tag{2.38}$$

$$\varphi = (I - L_0)^{-1}u_0.$$

We show that if condition (2.28) is satisfied with a sufficiently small r, then the operator Q is a contraction in the ball S_ρ of radius ρ with center at zero of the space H_2^T for any sufficiently small $\rho > 0$. Indeed, suppose u, $v \in S_\rho$. Using (2.36), we then obtain

$$\|Qu - Qv\|_{H_2^T} \leq \alpha\|u - v\|_{H_2^T};$$

$$\alpha = 2C_2C\rho; \qquad C_2 = \|(I - L_0)^{-1}\|_{H_2^T \to H_2^T}. \tag{2.39}$$

Suppose ρ is so small that $\alpha < 1$, while r satisfies

$$r \leq \alpha(2 - \alpha)/4CC_2^2. \tag{2.40}$$

The operator Q then takes S_ρ into itself. Indeed, applying (2.22) and (2.37), we obtain

$$\|Qu\|_{H_2^T} \leq C_2 r + C_2 C\rho^2 \leq \rho. \tag{2.41}$$

The existence of a solution $u \in H_2^T$ of the equation or of the Cauchy problem (1.1), (2.14) now follows from the principle of contraction mappings. This solution in S_ρ is unique. We shall show that uniqueness also holds in all of H_2^T. We suppose that there exist two solutions u_1, $u_2 \in H_2^T$ of problem (1.1), (2.14) with the same a.

Taking the inner product of v in H with the equation for their difference $v = u_1 - u_2$ (which follows from (1.1)) and integrating with respect to time, we obtain

$$\frac{1}{2}\|v(t)\|_H^2 + \int_0^t \|v(\tau)\|_{H_1}^2\, d\tau = \int_0^t \int_\Omega v \cdot (v, \nabla)(v_0 + u_1)\, dx\, d\tau. \tag{2.42}$$

for any $t \in [0, T]$. Since $v \in \widetilde{H}_2^T$ and $u \in H_2^T$, according to Lemma 2.2,

$$\int_0^T \|v_0 + u_1\|_{W_3^{(1)}}^4\, d\tau < \infty. \tag{2.43}$$

Therefore, applying the Hölder inequality and the imbedding theorem for H_1 in L_6 to estimate the right side of (2.42), we find that it does not exceed

$$C\int_0^t \|v(\tau)\|_H \cdot \|v(\tau)\|_{H_1}\,d\tau \le \int_0^t \|v(\tau\|^2_{H_1^2}\,d\tau + \frac{C^2}{4}\int_0^t \|v(\tau)\|^2_H\,d\tau, \quad (2.44)$$

where C is a known constant. From (2.42) and (2.44) we get

$$\|v(t)\|^2_H \le \int_0^t \|v(\tau)\|^2_H\,d\tau, \quad (2.45)$$

which, as we know, is satisfied only by $v(t) \equiv 0$.

Finally, applying (2.38), (2.22), and (2.37), we conclude that conditions (2.28) and (2.40) imply

$$\|u\|_{H_2^T} = \|Qu\|_{H_2^T} \le C_2\|a\|_{H_1} + \tfrac{1}{2}\alpha\|u\|_{H_2^T}, \quad (2.46)$$

from which (2.29) follows. Lemma 2.4 is thus proved.

We introduce the translation operator along trajectories of equation (1.1) by setting for $a \in H_1$

$$N_t a = u(t), \quad (2.47)$$

where $u(t)$ is the solution of the Cauchy problem (1.1), (2.14), and similarly, the translation operator along trajectories of the linear equation (1.3)

$$U_t a = \tilde{u}(t), \quad (2.48)$$

where $\tilde{u}$ is the solution of the Cauchy problem (1.3), (2.14).

According to Lemma 2.3, the linear operator U_t for fixed $t > 0$ is continuous in H_1. It follows from Lemma 2.4 that the nonlinear operator N_t is defined for $0 \le t \le T$ in some neighborhood of zero in H_1.

LEMMA 2.5. *The operator N_t $(0 \le t \le T)$ is defined in a neighborhood of zero of the space H_1, is compact in H_1, and is continuously differentiable. Its Fréchet differential at 0 is U_t. The operator U_t is also compact in H_1. In a neighborhood of zero in H_1 the operator N_t has Fréchet derivatives of any order.*

PROOF. Suppose $\|a\|_{H_1}$ is so small that the vector-valued function $N_t a = u(t)$ is defined for $0 \le t \le T$. We consider the equation

$$z = z_0 + M_0^0(z, u + v_0) \equiv z_0 + L_1 z; \quad (2.49)$$
$$z_0 = e^{-tA_0}h; \qquad M_0^0(u, z) \equiv M_0(u, z) + M_0(z, u),$$

where h is an arbitrary vector in H_1. This equation is uniquely solvable in H_2^T: indeed, it can be obtained from (2.20) by the change $a \to h$,

$v_0(t) \to v_0(t) + u(t)$, and hence the required result follows from Lemma 2.3. We set

$$z(t) = N_t'(a)h. \tag{2.50}$$

From the considerations of Lemma 2.3 we obtain the boundedness of the operaotr $N_t(a)$ in H_1 and, moreover, the uniform estimate

$$\|N_t'(a)\| \le C, \tag{2.51}$$

where the constant C does not depend on $t \in [0, T]$ and a if a varies in a small neighborhood of zero in H_1.

We shall show that $N_t(a)$ is the Fréchet differential of the operator N_t at the point a. For this it suffices to prove that for small $h \in H_1$

$$\|N_t(a+h) - N_t(a) - N_t'(a)h\|_{H_2^T} \le C\|h\|_{H_1}^2. \tag{2.52}$$

We introduce the notation

$$u_h(t) = N_t(a+h); \quad u(t) = N_t a; \quad z_h = u_h - u; \quad \psi = z_h - z.$$

From (2.30) and (2.49) we obtain

$$\psi = L_1\psi + M_0(z_h, z_h); \tag{2.53}$$

$$z_h = z_0 + L_1 z_h + M_0(z_h, z_h). \tag{2.54}$$

For the vector-valued function z_h for small $h \in H_1$ there is an estimate analogous to (2.29):

$$\|z_h\|_{H_2^T} \le C\|h\|_{H_1}. \tag{2.55}$$

From (2.53), using the boundedness of the operator $(I - L_1)^{-1}$ and the estimates (2.37) and (2.55), we derive

$$\|\psi(t)\|_{H_1} \le \|\psi\|_{H_2^T} \le C\|h\|_{H_1}^2, \tag{2.56}$$

which coincides with (2.51). Thus, $N_t'(a)$ is the Fréchet differential of the operator N_t at a. Since $N_t(0) = 0$, (2.49) for $a = 0$ coincides with (2.20); therefore, $N_t'(0) = U_t$.

In order to prove continuity of the operator-valued function $N_t'(a)$ we note that the vector-valued function $\Theta(t) = [N_t'(a+h) - N_t'(a)]g$ for any $g \in H_1$ and small $h \in H_1$ by (2.49) satisfies

$$\Theta = L_1\Theta + M_0^0(N_t'(a+h)g, z_h). \tag{2.57}$$

Considering the boundedness of the operator $(I - L_1)^{-1}$ and applying (2.34) and (2.51), from (2.57) we obtain

$$\|\Theta(t)\|_{H_1} \le \|\Theta\|_{H_2^T} \le C\|g\|_{H_1} \cdot \|h\|_{H_1}. \tag{2.58}$$

From (2.58) we obtain

$$\|N_t'(a+h) - N_t'(a)\|_{H_1 \to H_1} \le C\|h\|_{H_1}. \qquad (2.59)$$

Thus, the operator-valued function $N_t'(a)$ for any fixed $t \in [0, T]$ is continuous in t in a neighborhood of zero in H_1; the existence of a Fréchet derivative of any order can be proved similarly; here, by the way, it suffices to appeal to the implicit function theorem.

The continuity of N_t in H_1 follows from (2.52). We prove that it is compact. We introduce the vector-valued function $w(t) = \psi(t)u(t)$, where ψ is an infinitely differentiable function equal to 0 for $0 \le t \le \delta/2$ and to 1 for $t \ge \delta$; δ is any number in $(0, T)$. From (1.1) we obtain

$$\frac{dw}{dt} + A_0 w = \psi(Bu + Ku) + \psi' u \equiv q; \qquad w(0) = 0. \qquad (2.60)$$

Using Lemma 2.2 and the fact that $v_0 \in \widetilde{H}_2^T$, for the right side q of (2.60) we obtain

$$\left[\int_0^t \int_\Omega |q(x,t)|^{5/2}\, dx\, dt\right]^{2/5} \le C_1\|u\|_{H_2^T} + C_2\|u\|^2_{H_2^T}. \qquad (2.61)$$

According to Theorem 7.3 of Chapter I, from (2.60) we obtain

$$\int_0^t \int_\Omega (|D_t w|^{5/2} + |D_x^2 w|^{5/2})\, dx\, dt \le \int_0^T \int_\Omega |q|^{5/2}\, dx\, dt. \qquad (2.62)$$

Recalling that $w(t) = u(t)$ for $t \ge \delta$ and considering (2.29), from (2.61) and (2.62) we deduce the inequality

$$\int_0^T \int_\Omega (|D_t u|^{5/2} + |D_x^2 u|^{5/2})\, dx\, dt \le C_\delta(\|a\|^{5/2}_{H_1} + \|a\|^5_{H_1}). \qquad (2.63)$$

From (2.63) it follows that the operator N_t maps a neighborhood of zero of the space H_1 into a bounded set of the Nikol'skiĭ space

$$H_1^{5/2-\varepsilon, 5/2-\varepsilon}(\Omega \times [\delta, T]), \qquad \varepsilon > 0,$$

which is compactly imbedded in $W_2^{(1)}$ and hence in H_1 (see [59]). Therefore, the operator N_t is compact. Compactness of U_t can be proved similarly. Lemma 2.5 is thus proved.

We remark that, as is not hard to show, the operator N_t is not only differentiable but also analytic in a for any $t \in [0, T]$. Its complete continuity could be proved without using Nikol'skiĭ's result by establishing that under the conditions of Lemmas 2.3 and 2.4 the successive approximations converge in $W_2^{(2)}(\Omega)$ for any t.

We further note a useful corollary of Lemma 2.5.

COROLLARY. *The spectrum of the monodromy operator U_t contains no more than a finite number of eigenvalues greater than 1 in modulus. The stability spectrum contains no more than a finite number of eigenvalues with positive and distinct real parts.*

PROOF. The first assertion is a consequence of the compactness of U_t. To prove the second it suffices to note that if σ is an eigenvalue of problem (1.5), then $\rho = e^{T\sigma}$ is an eigenvalue of U_T to which there corresponds the eigenfunction $w(0)$.

§3. A condition for asymptotic stability

We first consider in an arbitrary real([13]) Banach space X the equation

$$x = Nx, \tag{3.1}$$

where N is an operator defined in some neighborhood D_0 of zero of the space X which is continuous and continuously differentiable in this neighborhood. Suppose, moreover, that $N0 = 0$, so that the point 0 is a solution of (3.1). We denote by $N'(a)$ the Fréchet differential of the operator N at the point $a \in D_0$, and we set $N'(0) = U$. The following assertion holds.

LEMMA 3.1. *Suppose the spectrum of the operator U is contained inside the unit disk,*

$$|\sigma(U)| \leq \rho < 1. \tag{3.2}$$

Then the successive approximations defined by

$$x_{n+1} = Nx_n, \tag{3.3}$$

converge to 0 in norm for any x_0 in a neighborhood $D \subset D_0$ of zero of the space X. If $\rho_0 > \rho$, then there exists a constant C, depending only on ρ_0 and the operator N, such that

$$\|N^n x_0\| \leq C\rho_0^n \|x_0\|. \tag{3.4}$$

PROOF. Let ρ_0 be any number between ρ and 1. For the operator U^n $(n = 1, 2, \dots)$ there is then the representation

$$U^n = \frac{1}{2\pi i} \int_{|\lambda|=\rho_0} \lambda^n (\lambda I - U)^{-1}\, d\lambda, \tag{3.5}$$

from which we obtain

$$\|U^n\|_{X\to X} \leq B\rho_0^n; \qquad B = \max_{|\lambda|=\rho_0} \|(\lambda I - U)^{-1}\|. \tag{3.6}$$

([13])A linear operator A acting in X is henceforth considered to be extended in the natural way to the complex hull of the space X. Here any invariant subspace of the extended operator containing together with any element x its complex conjugate x^* is the complex hull of some invariant subspace of A in X. Below only such invariant subspaces are considered.

Therefore, $\|U^n\| \to 0$ as $n \to \infty$, and it is possible to choose a natural number k such that

$$\|U^k\| \leq B\rho_0^k < 1. \tag{3.7}$$

The operator N^k is then a contraction in any ball $D_r = \{x : \|x\| \leq r\} \subset D_0$ of sufficiently small radius r. Indeed, its Fréchet differential at a point $x \in D_0$ under the condition that $Nx, N^2x, \dots, N^kx \in D_0$ is

$$(N^k)'(x) = N'(N^{k-1}x) \cdot N'(N^{k-2}x) \cdots N'(Nx) \cdot N'(x), \tag{3.8}$$

which for $x = 0$ coincides with U^k. Since the operator-valued function (3.8) is continuous in a neighborhood of $x = 0$, for sufficiently small r we have for all $x \in D_r$

$$\|(N^k)'(x)\|_{X \to X} \leq B\rho_0^k < 1. \tag{3.9}$$

If $x,\ x + h \in D_r$, then, using (3.9), we obtain

$$\|N^k(x+h) - N^kx\|_X = \left\| \int_0^1 (N^k)'(x+th)h\,dt \right\| \leq q\|h\|_X; \tag{3.10}$$

$$\|N^kx\| = \left\| \int_0^1 (N^k)'(tx)x\,dt \right\| \leq q\|x\|, \tag{3.11}$$

$$q = B\rho_0^k < 1.$$

Thus, N^k is a contraction in the ball D_r. Therefore 0 is the unique solution of the equation $x = N^kx$ and thus of (3.1) in D_r. It remains to show that (3.4) holds. For this, setting $n = \alpha k + \beta$, where α and β are natural numbers, $0 \leq \beta \leq k-1$, and applying (3.11), for any $x_0 \in D_r$ we obtain

$$\|N^nx_0\| = \|N^{\alpha k}(N^\beta x_0)\| \leq q^\alpha \|N^\beta x_0\|. \tag{3.12}$$

Decreasing r if necessary, we can assume that the functional $\|Nx\|$ is bounded on the set $\bigcup_{\beta=1}^{k-1} N^\beta(D_r)$, say, by a constant C_1. For any $x_0 \in D_r$ we then have

$$\|N^\beta x_0\| = \left\| \int_0^1 N'(N^{\beta-1}(tx_0)) \cdots N'(N(tx_0))N'(tx_0)x_0\,dt \right\|$$
$$\leq C_1^\beta \|x_0\|. \tag{3.13}$$

comparing (3.12) with (3.13) and noting that $q = B\rho_0^k$, we arrive at (3.4). Lemma 3.1 is proved.

THEOREM 3.1. *Suppose the stability spectrum of the periodic flow $v_0(t)$ lies within the left half-plane, i.e., for all eigenvalues σ of problem* (1.5)

$$\operatorname{Re}\sigma < -\sigma_0 < 0. \tag{3.14}$$

Then the flow v_0 is asymptotically Lyapunov stable in H_1. Moreover, for any solution of the Cauchy problem (1.1), (2.14), *for sufficiently small $\|a\|_{H_1}$,*

$$\|u(t)\|_{H_1} \le Ce^{-\sigma_0 t}\|a\|_{H_1}; \tag{3.15}$$

$$\int_0^t e^{2\sigma_0\tau}\left(\left\|\frac{du}{d\tau}\right\|_H^2 + \|A_0 u\|\|_H^2\right)d\tau \le C^2\|a\|_{H_1}^2. \tag{3.16}$$

PROOF. Since by Lemma 2.5 the monodromy operator U_T is compact in H_1, its spectrum consists of the point 0 and eigenvalues $\rho_1, \rho_2, \dots$, of finite multiplicity, which are called *multipliers*. There is a simple connection between multipliers and points of the stability spectrum. Namely, if σ is an eigenvalue of problem (1.5), then $\rho = e^{T\sigma}$ is a multiplier; if ρ is a multiplier, then the numbers $\sigma_k = (1/T)(\ln\rho + 2k\pi i)$, $k = 0, 1, \dots$ (the branch of the logarithm is chosen arbitrarily), belong to the stability spectrum. Indeed, if $w \in H_2^T$ is an eigenvector of problem (1.5), then $e^{\sigma t}w(t)$ is a solution of the Cauchy problem (1.3), (2.14) with initial vector $a = w(0)$. Therefore, $U_T a = \rho a$. Conversely, if the last equation is satisfied and $u(t)$ is a solution of the Cauchy problem (1.3), (2.14), then the vector-valued function $w_k(t) = e^{-\sigma_k t}u(t)$ is T-periodic in t and satisfies (1.5) for $\sigma = \sigma_k$.

Since $|\rho| = e^{T\operatorname{Re}\sigma}$, it follows from the hypothesis of the theorem that the spectrum of the monodromy operator is contained within the unit disk:

$$|\sigma(U)| < \rho_0 = e^{-\sigma_0 T} < 1. \tag{3.17}$$

From Lemmas 2.5 and 3.1 it now follows that in the space H_1 there exists a neighborhood of zero $D_r = \{a \in H_1 : \|a\|_{H_1} < r\}$ such that the nonlinear Cauchy problem (1.1), (2.14) for any $a \in D_r$ has a solution $u(t)$ defined for all $t > 0$, and $\|u(nT)\|_{H_1} = \|U_T^n a\|_{H_1} \to 0$ as $n \to \infty$. Applying (2.29), we find that $\|u(t)\|_{H_1} \to 0$ as $t \to +\infty$. Thus, asymptotic stability of the flow $v_0(t)$ in H_1 is proved.

According to Lemma 3.1, from condition (3.17) we obtain

$$\|u(nT)\|_{H_1} \le C_1 e^{-n\sigma_0 T}\|a\|_{H_1}; \qquad (n = 1, 2, \dots). \tag{3.18}$$

If $nT \le t < (n+1)T$, from (2.29) and (3.18) we deduce that

$$\|u(t)\|_{H_1} \le C_0\|u(nT)\|_{H_1} \le C_2 e^{-n\sigma_0 T}\|a\|_{H_1} \le Ce^{-\sigma_0 t}\|a\|_{H_1};$$
$$C - C_2 e^{\sigma_0 t}; \tag{3.19}$$

$$\int_{nT}^{t}\left(\left\|\frac{du}{d\tau}\right\|_{H}^{2}+\|A_0u(\tau)\|_H^2\right)d\tau\leq C^2e^{-2\sigma_1 t}\|a\|_{H_1}^2, \tag{3.20}$$

where σ_1 is any number such that $\sigma_1 > \sigma_0$ but $\operatorname{Re}\sigma < -\sigma_1$ for all eigenvalues σ. The estimate (3.19) coincides with (3.15). Inequality (3.16) follows immediately from (3.20). Indeed, introducing the notation $\psi^2(t) = \|du/dt\|_H^2 + \|A_0u\|_H^2$, we have

$$\begin{aligned}\int_0^t e^{2\sigma_0\tau}\psi^2(\tau)\,d\tau &\leq \sum_{k=1}^{n+1}\int_{(k-1)T}^{kT} e^{2\sigma_0\tau}\psi^2(\tau)\,d\tau\\ &\leq C^2e^{2\sigma_1 T}\sum_{k=1}^{\infty}e^{-2(\sigma_1-\sigma_0)kT}\|a\|_{H_1}^2.\end{aligned} \tag{3.21}$$

Since the series in (3.21) converges, (3.16) follows from this inequality. Theorem 3.1 is proved.

§4. A condition for instability

We again consider equation (3.1) and prove the following auxiliary assertion.

LEMMA 4.1. *Suppose that the spectrum of the operator U is the union of nonintersecting closed sets $\sigma_1(U)$ and $\sigma_2(U)$, and*

$$|\sigma_1(U)| > \beta_1 > 1; \qquad |\sigma_2(U)| \leq 1. \tag{4.1}$$

Then there exists $\varepsilon_0 > 0$ such that for any $\delta > 0$ it is possible to find a vector $a \in X$, $\|a\| < \delta$, such that

$$|N^{n_0}a\| \geq \varepsilon_0 \tag{4.2}$$

for some natural number n_0.

PROOF. Suppose the lemma is false. Then for any $\varepsilon > 0$ it is possible to find $\delta > 0$ such that

$$\|N^n a\| < \varepsilon, \qquad (n = 0, 1, 2, \dots), \tag{4.3}$$

provided that $\|a\| < \delta$. We set

$$Na = Ua + Ra. \tag{4.4}$$

Since U is the Fréchet differential of the operator N at 0, for any $\Theta > 0$ it is possible to choose $\varepsilon > 0$ so small that $\|a\| < \varepsilon$ implies

$$\|Ra\| \leq \Theta\|a\|. \tag{4.5}$$

We further denote by P_1 and P_2 the projections corresponding to the spectral sets $\sigma_1(U)$ and $\sigma_2(U)$:

$$P_1 = \frac{1}{2\pi i}\int_{\gamma_1}(\lambda I - U)^{-1}\,d\lambda; \qquad P_2 = \frac{1}{2\pi i}\int_{\gamma_2}(\lambda I - U)^{-1}\,d\lambda, \tag{4.6}$$

where the smooth contours γ_1 and γ_2, respectively, encompass the sets $\sigma_1(U)$ and $\sigma_2(U)$, where γ_1 lies entirely outside and γ_2 entirely inside the closure of the unit disk of the λ plane. Here $P_1 + P_2 = I$ and $P_1 P_2 = P_2 P_1 = 0$. The space X decomposes into a direct sum of the subspaces $X_1 = P_1(X)$ and $X_2 = P_2(X)$, which are invariant relative to the operator U. The spectrum of the restriction U_k of U to X_k is $\sigma_k(U)$, $k = 1, 2$.

We remark that for any $a' \in X_1$

$$\|U_1^n a'\| \geq M_1 \beta_1^n \|a'\| \qquad (n = 1, 2, \dots), \tag{4.7}$$

where the constant $M_1 > 0$ does not depend either on n or on a'; the constant β_1 satisfies (4.1). This estimate follows directly from the relation

$$a' = -\frac{1}{2\pi i} \int_{|\lambda| = \beta_1} \lambda^{-n} (\lambda I - U_1)^{-1} U_1^n a' \, d\lambda.$$

Here for the constant M_1 we obtain

$$\frac{1}{M_1} = \beta_1 \cdot \max_{|\lambda| = \beta_1} \|(\lambda I - U_1)^{-1}\|.$$

Suppose $1 < \beta_2 < \beta_1$. Then for the operator U_2^n there is the integral representation

$$U_2^n = \frac{1}{2\pi i} \int_{|\lambda| = \beta_2} \lambda^n (\lambda I - U_2)^{-1} \, d\lambda.$$

From it we derive the estimate

$$\|U_2^n\| \leq M_2 \beta_2^n; \qquad M_2 = \max_{|\lambda| = \beta_2} \|(\lambda I - U_2)^{-1}\|. \tag{4.8}$$

From (4.7) and (4.8) it follows that for sufficiently large m

$$\|U_2^m\| < M_1 \beta_1^m > 1. \tag{4.9}$$

It may be assumed that (4.9) is satisfied already for $m = 1$, since we can always go over to this case by replacing the operator N by N^m. We thus assume that

$$\|U_2\| < M_1 \beta_1 > 1. \tag{4.10}$$

Suppose $k > 0$ is any number. We choose ε in (4.3) so small that the constant Θ in (4.5) satisfies

$$\Theta < \frac{(M_1 \beta_1 - 1)k}{(k+1)\|P_2\|}; \qquad \Theta < \frac{k(M_1 \beta_1 - \|U_2\|)}{(1+k)(k\|P_2\| + \|P_2\|)};$$
$$\Theta < \frac{M_1 \beta_1}{\|P_1\| + k\|P_2\|}. \tag{4.11}$$

In the space X we consider the set $\Sigma_{k,\varepsilon}$ consisting of points $a \in X$ satisfying

$$\|a\| < \varepsilon; \qquad \|P_1 a\| \geq k\|P_2 a\|. \tag{4.12}$$

We show that assumption (4.3) entails the estimate

$$\|P_1 N^n a\| \geq l^n \|P_1 a\|; \qquad l = M_1\beta_1 - \Theta\|P_1\|(1 + 1/k). \tag{4.13}$$

Here $l > 1$ by the first inequality of (4.11).

Suppose $a \in \Sigma_{k,\varepsilon}$. Then $\|Na\| < \varepsilon$ according to (4.3).

Further, from (4.4) and (4.7) for $n = 1$ we conclude that

$$\begin{aligned} \|P_1 Na\| &= \|U_1 P_1 a + P_1 Ra\| \geq M_1\beta_1\|P_1 a\| - \Theta\|P_1\| \cdot \|a\|; \\ \|P_2 Na\| &= \|U_2 P_2 a + P_2 Ra\| \leq \|U_2\| \cdot \|P_2 a\| + \Theta\|P_2\| \cdot \|a\|. \end{aligned} \tag{4.14}$$

Applying (4.12), (4.11), and the elementary inequality $\|a\| \leq \|P_1 a\| + \|P_2 a\|$, from (4.14) we conclude that

$$\|P_1 Na\| \geq k\|P_2 Na\|. \tag{4.15}$$

Thus, $Na \in \Sigma_{k,\varepsilon}$. Then also $N^n a \in \Sigma_{k,\varepsilon}$ $(n = 1, 2, \ldots)$. We prove that (4.13) is satisfied. We use induction. For $n = 1$ we obtain (4.13) by estimating the right side of the first inequality of (4.14) by means of (4.12);

$$\|P_1 Na\| \geq M_1\beta_1\|P_1 a\| - \Theta\|P_1\|(\|P_1 a\| + \|P_2 a\|) \geq l\|P_1 a\|. \tag{4.16}$$

If we suppose that (4.13) is satisfied for some value of n for any $a \in \Sigma_{k,\varepsilon}$, then, noting that $Na \in \Sigma_{k,\varepsilon}$ and applying (4.16), we find that (we emphasize that in the derivation of (4.16) only the second inequality of (4.12) is used)

$$\|P_1 N^{n+1} a\| = \|P_1 N^n(Na)\| \geq l^n\|P_1 Na\| \geq l^{n+1}\|P_1 a\|. \tag{4.17}$$

Thus, (4.13) is proved for any $n = 1, 2, \ldots$. From this inequality it follows that $\|P_1 N^n a\| \to \infty$ for any $a \in \Sigma_{k,\delta}$ provided that $a \neq 0$. But this contradicts the assumption (4.3), by which

$$\|P_1 N^n a\| \leq \varepsilon\|P_1\|. \tag{4.18}$$

Lemma 4.1 is thus proved.

THEOREM 4.1. *Suppose the stability spectrum of the periodic flow $v_0(t)$ contains at least one eigenvalue σ_0 with positive real part. Then the flow v_0 is unstable in H_1.*

PROOF. Since the monodromy operator is completely continuous, the set $\sigma_1(U_T)$ of its eigenvalues larger than 1 in modulus is not more than finite. By the hypothesis of the theorem it is nonempty, since it contains the eigenvalue $\rho_0 = e^{\sigma_0 T}$ and $|\rho_0| > 1$. Therefore, Theorem 4.1 follows immediately from Lemmas 4.1 and 2.5.

§5. Conditional stability

We return again to equation (3.1).

LEMMA 5.1. *Suppose that the spectrum of the operator U can be represented as the union of nonintersecting closed sets $\sigma_1(U)$ and $\sigma_2(U)$, where*

$$|\sigma_1(U)| > 1; \qquad |\sigma_2(U)| < 1. \tag{5.1}$$

Then in some neighborhood $D_r = \{x \in X : \|x\| < r\}$ of zero of the space X there are manifolds Y_1 and Y_2, invariant relative to the operator N, which at zero are tangent respectively to the space X_1 and X_2 (defined in Lemma 4.1). Here 1) if $x_0 \in Y_2$, then the successive approximations $N^n x_0$ as $n \to \infty$ converge to 0; 2) if $x_0 \notin Y_2$, then $\|N^n x_0\| > r$ for some n; 3) for any $x_0 \in Y_1$ the inverse mapping $N^{-1}x_0 \in Y_1$ is defined, and $N^{-n}x_0 \to 0$ as $n \to \infty$; and 4) if $x_0 \notin Y_1$, then for some n either the elements $N^{-n}x_0$ are not defined or $\|N^{-n}x_0\| > r$.

PROOF. The space X decomposes into a direct sum of the subspaces X_1 and X_2 as in Lemma 4.1. We write an element $x \in X$ in the form $x = (x_1, x_2)$ where $x_1 = P_1 x$ and $x_2 = P_2 x$. The operator N acts on a point x according to the law

$$\begin{aligned} x_1 &\to U_1 x_1 + f_1(x_1, x_2); \\ x_2 &\to U_2 x_2 + f_2(x_1, x_2), \end{aligned} \tag{5.2}$$

where $f_k(x_1, x_2) = P_k N x - U_k x_k$, $k = 1, 2$. We seek the manifold Y_2 in the form

$$x_1 = \varphi(x_2), \tag{5.3}$$

where $\varphi : X_2 \to X_1$ is a continuous operator defined in the neighborhood $D_r^2 = \{x_2 \in X_2 : \|x_2\| \le r > 0\}$ of zero of the space X_2. The condition that the manifold (5.3) be invariant relative to the transformation (5.2) can be written in the form

$$\begin{aligned} \varphi(x_2) &= U_1^{-1}\varphi(U_2 x_2 + f_2(\varphi(x_2), x_2)) - U_1^{-1} f_1(\varphi(x_2), x_2) \\ &\equiv (L\varphi)(x_2), \qquad x_2 \in D_r^2. \end{aligned} \tag{5.4}$$

We introduce the Banach space $C(D_r^2 \to X_1) = C$ of mappings of D_r into X_1 which are bounded and uniformly continuous. The norm in it is defined by

$$\|\varphi\| = \sup_{y \in D_r^2} \|\varphi(y)\|_{X_1}. \tag{5.5}$$

In it we consider the set $\omega(r, \rho, \gamma) = \omega$ consisting of those $\varphi \in C(D_r^2 \to X_1)$ for which $\varphi(0) = 0$ and([14])

$$\|\varphi\| \leq \rho; \qquad \|\varphi(y_1) - \varphi(y_2)\|_{X_1} \leq \gamma \|y_1 - y_2\|_{X_2}; \\ (y_1, y_2 \in D_r^2). \tag{5.6}$$

It is easy to see that $\omega(r, \rho, \gamma)$ is a closed set and is hence a complete metric space.

We shall show that if the positive numbers r, ρ, and γ are sufficiently small, then the operator L defined in (5.4) acts in $\omega(r, \rho, \gamma)$ and some power of it is a contraction. We first consider the operator $N_\varphi : X_2 \to X_2$ defined for a given $\varphi \in \omega(r, \rho, \gamma)$ by

$$N_\varphi x_2 = U_2 x_2 + f_2(\varphi(x_2), x_2). \tag{5.7}$$

It may be assumed with no loss of generality that the operators U_1^{-1} and U_2 are contractions. Indeed, this can be arranged by introducing in X the new norm, equivalent to the previous one,

$$\|x\|_* = \sum_{k=1}^{\infty} \|U_1^{-k} P_1 x\| + \sum_{k=0}^{\infty} \|U_2^k P_2 x\|. \tag{5.8}$$

The equivalence of the norms follows from the inequalities

$$2\|x\| \leq \|x\|_* \leq \left(\sum_{k=0}^{\infty} \|U_1^{-k}\| \cdot \|P_1\| + \sum_{k=0}^{\infty} \|U_2^k\| \cdot \|P_2\| \right) \|x\|. \tag{5.9}$$

Convergence of the series is guaranteed by condition (5.1).

The fact that the norms of U^{-1} and U_2 are less than 1 follows from (5.9) and the relations, following from (5.8),

$$\|U_2 P_2 x\|_* = \|P_2 x\|_* - \|P_2 x\|; \qquad \|U_1^{-1} P_1 x\| = \|P_1 x\|_* - \|P_1 x\|. \tag{5.10}$$

We thus assume that the following conditions are satisfied:

$$\|U_1^{-1}\| = q_1 < 1; \qquad \|U_2\| = q_2 < 1. \tag{5.11}$$

From (5.11) it now follows that the operator N_φ takes D_r^2 into itself if r and ρ are sufficiently small. Inded, from (5.7) we conclude that for $x \in D_r^2$

$$\|N_\varphi x_2\|_{X_2} \leq q_2 r + o(\rho + r) < r. \tag{5.12}$$

Thus, the vector-valued function $L\varphi$ is well-defined for all $\varphi \in \omega(r, \rho, \gamma)$. Further, from (5.4) with the use of (5.11) for any $\varphi \in \omega(r, \rho, \gamma)$ for small r, ρ, and γ we obtain

$$\|L\varphi\|_C \leq q_1 \rho + o(\rho + r) < \rho; \tag{5.13}$$

([14])The condition $\varphi(0) = 0$ can be dropped and then derived from (5.4).

$$\|(L\varphi)(y_1) - (L\varphi)(y_2)\|_{x_1} \leq \gamma_1 \|y_1 - y_2\|. \tag{5.14}$$

Here for the constant γ_1 we have

$$\begin{gathered}\gamma_1 = q_1 q_2 \gamma + q_1 \gamma^2 m_{21}(r, \rho) + q_1 \gamma m_{22}(r, \rho) \\ + q_1 \gamma m_{11}(r, \rho) + q_1 m_{12}(r, \rho); \\ m_{ik}(r, \rho) = \sup_{\substack{\|x_1\| \leq \rho \\ \|x_2\| \leq r}} \left\| \frac{\partial f_i(x_1, x_2)}{\partial x_k} \right\|\end{gathered}$$

Since the operator-valued functions $\partial f_i(x_1, x_2)/\partial x_k$ become zero for $x_1 = x_2 = 0$ and are continuous, $m_{ik}(r, \rho) \to 0$ as $r, \rho \to 0$. Therefore, for sufficiently small r, ρ, and γ it follows from (5.11) that $\gamma_1 < \gamma$.

Further, for any $\varphi_1, \varphi_2 \in \omega(r, \rho, \gamma)$ we have

$$\begin{gathered}\|L\varphi_1 - L\varphi_2\|_C \leq \alpha \|\varphi_1 - \varphi_2\|_C; \\ \alpha = q_1 + q_1 \gamma m_{21}(r, \rho) + q_1 m_{11}(r, \rho); \qquad \alpha < 1\end{gathered} \tag{5.15}$$

for sufficiently small r, ρ, and γ.

Inequality (5.15) shows that the operator L in $\omega(r, \rho, \gamma)$ is a contraction. Therefore, in $\omega(r, \rho, \gamma)$ equation (5.4) has a unique solution φ. Thus, in a neighborhood of the point $0 \in X$ equation (5.3) defines a set Y_2 invariant relative to the mapping (5.2). Clearly this is a manifold (a homeomorphic image of the ball D_r^2 of the space X_2). Indeed, the mapping $(0, x_2) \to (\varphi(x_2), x_2)$ provides the homeomorphism. The manifold Y_2 is tangent to the point $0 \in X$ of the subspace X_2, i.e., $\|\varphi(x_2)\| = o(\|x_2\|)$ for $\|x_2\| \to 0$. We prove this. From (5.4), setting $\psi(x_2) = \varphi(x_2) - U^{-1}\varphi(U_2 X_2)$ ($x_2 \in D_R^1$) and using (5.6) and the fact that $\varphi(0) = 0$, for any $\varepsilon > 0$ we obtain

$$\begin{aligned}\|\psi(x_2)\| &\leq q_1 \gamma \|f_2(\varphi(x_2), x_2)\| + q_1 \|f_1(\varphi(x_2), x_2)\| \\ &\leq \varepsilon(1 - q_1)\|x_2\|\end{aligned} \tag{5.16}$$

for all sufficiently small x_2: $\|x_2\| \leq r_\varepsilon$. For such x_2, expressing φ in terms of ψ and estimating with the help of (5.11) and (5.16), we obtain

$$\begin{aligned}\|\varphi(x_2)\|_{x_1} &= \left\| \sum_{k=0}^{\infty} U_1^{-k} \psi(U_2^k x_2) \right\| \\ &\leq \sum_{k=0}^{\infty} q_1^k \cdot \varepsilon(1 - q_1)\|x_2\| \leq \varepsilon \|x_2\|,\end{aligned} \tag{5.17}$$

which was required.

Suppose now that $D_0 = \{x \in X : \|x\| \leq \varepsilon_0\}$ is a ball with center at $0 \in X$ of sufficiently small radius ε_0. We show that for any $x_0 \in D_r - Y_2$ we have, for sufficiently large n,

$$\|N^n x_0\| \geq \varepsilon_0. \tag{5.18}$$

We introduce the notation

$$x_0 = (x_1^0, x_2^0); \quad N^n x_0 = (x_1^n, x_2^n); \quad n = 0, 1, \ldots; \quad x_1^n \in X_1; \quad x_2^n \in X_2.$$

Suppose that $d_n = \|x_1^n - \varphi(x_2^n)\|$. We suppose that

$$\|N^n x_0\| < \varepsilon_0. \tag{5.19}$$

We estimate d_{n+1}. Applying (5.2) and (5.4), we obtain

$$\begin{aligned} d_{n+1} \geq \|U_1(x_1^n - \varphi(x_2^n))\| &- \|f_1(x_1^n, x_2^n) - f_1(\varphi(x_2^n), x_2^n)\| \\ &- \|\varphi(U_2 x_2^n + f_2(x_1^n, x_2^n)) - \varphi(U_2 x_2^n - f_2(\varphi(x_2^n), x_2^n))\|. \end{aligned} \tag{5.20}$$

Further, using (5.11) and (5.6), we conclude that

$$d_{n+1} \geq p d_n; \qquad p = 1/q_1 - m_{11}(\varepsilon_0) - \gamma m_{21}(\varepsilon_0); \tag{5.21}$$

$$m_{ik}(\varepsilon_0) = \sup_{x \in D_0} \left\| \frac{\partial f_i}{\partial x_k} \right\|.$$

Since $q_1 < 1$ and $m_{ik}(\varepsilon_0) \to 0$ as $\varepsilon_0 \to 0$, for sufficiently small ε_0 we find that $p > 1$. Therefore, from (5.21) we obtain

$$d_n \geq p^n d_0; \qquad p > 1. \tag{5.22}$$

Now if (5.19) were satisfied for all n, then d_n would be bounded:

$$d_n \leq \sup_{x \in D_0} \|x_1 - \varphi(x_2)\| \leq (\|P_1\| + \gamma \|P_2\|)\varepsilon_0. \tag{5.23}$$

Thus, if $x \notin Y_2$, then there is an n such that $N^n x_0 \notin D_0$. Finally, we shall show that $N^n x_0 \to 0$ if $x_0 \in Y_2$. According to (5.2), (5.4), and (5.6), under condition (5.19) we have

$$\|x_2^{n+1}\| \leq p_0 \|x_2^n\|; \qquad p_0 = q_2 + \gamma m_{21}(\varepsilon_0) + m_{22}(\varepsilon_0). \tag{5.24}$$

Since $p_0 \to q_2$ as $\varepsilon_0 \to 0$, it may be assumed that $p_0 < 1$ for sufficiently small ε_0. From (5.24) it then follows that $x_2^n \to 0$ as $n \to \infty$. Hence, also $x_1^n = \varphi(x_2^n) \to 0$. Since $\|x^n\| \leq \|x_1^n\| + \|x_2^n\|$, we obtain $\|x^n\| \to 0$ $(n \to \infty)$.

Thus, all assertions of Lemma 5.1 regarding the manifold Y_2 are proved. If the operator U_2 is invertible, then the arguments for Y_1 are entirely similar. However, this condition is not satisfied in the most interesting case where the operator U_2 is compact and the space X is infinite-dimensional. The proof must therefore be carried out anew.

Thus, we seek a manifold Y_1 given by

$$x_2 = \psi(x_1), \tag{5.25}$$

where $\psi : X_1 \to X_2$ is a continuous operator defined in the ball $D_r^1 = \{x_1 \in X_1 : \|x_1\| \le r\}$. The condition of invariance of the manifold (5.25) relative to the transformation (5.2) is equivalent to the equality

$$\psi(U_1x_1 + f_1(x_1, \psi(x_1))) = U_2\psi(x_1) + f_2(x_1, \psi(x_1)). \tag{5.26}$$

We seek a solution ψ of (5.26) in the set $\omega_1(r, \rho, \gamma) = \omega_1$ of the space $C(D_r^1 \to X_2)$ consisting of those vector-valued functions ψ for which $\psi(0) = 0$ and

$$\|\psi\|_C \le \rho; \qquad \|\psi(x_1') - \psi(x_1'')\|_{X_2} \le \gamma\|x_1' - x_1''\|_{X_1}, \tag{5.27}$$
$$(x_1', x_1'' \in D_r^1).$$

In (5.26) we make the change of variable

$$x_1 \to z = U_1x_1 + f_1(x_1, \psi(x_1)) \equiv N_\psi x_1. \tag{5.28}$$

We note that N_ψ is a homeomorphic mapping of the ball D_r^1 onto some neighborhood of zero in X_1 if r, ρ, and γ are sufficiently small. Indeed, in order to find x_1 for a given z it is necessary to solve the equation

$$x_1 = U_1^{-1}z - U_1^{-1}f_1(x_1, \psi(x_1)), \tag{5.29}$$

whose right side for small z, r, ρ, and γ obviously defines a contraction operator in D_r^1. Equation (5.29) is thus uniquely solvable in D_r^1. Here for any given q, $(q_1 < q < 1)$ it is possible to choose r, ρ, and γ so that

$$\|x_1' - x_1''\| \le q\|N_\psi x_1' - N_\psi x_1''\|; \qquad x_1', x_1'' \in D_r^1. \tag{5.30}$$

Indeed, from (5.29) we obtain

$$\|x_1' - x_1''\| \le q_1\|z' - z''\| + \eta\|x_1' - x_1''\|; \tag{5.31}$$
$$\eta = q_1[m_{11}(r, \gamma r) + \gamma m_{12}(r, \gamma r)]; \quad z' = N_\psi x_1'; \quad z'' = N_\psi x_1''.$$

We obtain (5.30) from (5.31) by setting $q = q_1/(1 - \eta)$ and noting that $\eta \to 0$ as $r \to 0$; $\gamma < 1$.

After the change (5.28), (5.26) takes the form

$$\psi(z) = U_2\psi(N_\psi^{-1}z) + f_2(N_\psi^{-1}z, \psi(N_\psi^{-1}z)) \equiv (Q\psi)(z). \tag{5.32}$$

We show that the operator Q is a contraction in $\omega_1(r, \rho, \gamma)$ if (the reader please excuse us!) r, ρ, and γ are sufficiently small.

If $\psi \in \omega_1(r, \rho, \gamma)$, then, considering (5.30), we have

$$\|Q\psi\|_C \le q_2\rho + m_{21}(r, \gamma r) \cdot r + m_{22}(r, \gamma r)\rho \le \rho; \tag{5.33}$$

$$\|(Q\psi)(z_1) - (Q\psi)(z_2)\| \le \gamma_1\|z_1 - z_2\|; \tag{5.34}$$
$$\gamma_1 = q_2q\gamma + m_{21}(r, \gamma r)q + q\gamma m_{22}(r, \gamma r).$$

We fix γ $(0 < \gamma < 1)$ arbitrarily. We choose r so small that $q_2 q + q m_{22}(r,r) \leq \tau < 1$ and $m_{21}(r,r) < 1 - \tau$. Then $\gamma_1 < \gamma$ and the operator Q acts in ω_1.

Further, for any $\psi_1, \psi_2 \in \omega_1$, from (5.30) and (5.32) we conclude that

$$\|Q\psi_1 - Q\psi_2\|_C \leq \overline{q_2}\|\psi_1 - \psi_2\|_C + p_2 \max_{z \in D_r^1} \| - N_{\psi_2}^{-1} z\|; \tag{5.35}$$

$$\overline{q_2} = q_2 + m_{22}(qr, \gamma qr); \qquad p_2 = \gamma[q_2 + m_{22}(qr, \gamma qr)] + m_{21}(qr, \gamma qr).$$

To estimate the second term on the right side of (5.35) we again turn to (5.28). Setting $x' = N_{\psi_1}^{-1} z$ and $x'' = N_{\psi_2}^{-1} z$, we have

$$\|x' - x''\| \leq \|U_1^{-1}[f_1(x'', \psi_2(x'')) - f_1(x', \psi_1(x'))]\|. \tag{5.36}$$

Further,

$$\begin{aligned} \|x' - x''\| &\leq q_1 m_{11}(r, \gamma r)\|x' - x''\| \\ &\quad + q_1 m_{12}(r, \gamma r)[\|\psi_1 - \psi_2\|_C + \gamma\|x' - x''\|]. \end{aligned} \tag{5.37}$$

It follows from (5.37) that for fixed $\gamma < 1$ and arbitrary $\Theta > 0$ we can choose r so small that

$$\|x' - x''\| \leq \Theta\|\psi_1 - \psi_2\|_C. \tag{5.38}$$

From (5.35) and (5.37) it follows that for any q_3 such that $q_2 < q_3 < 1$ we can choose r and γ so that

$$\|Q\psi_1 - Q\psi_2\|_C \leq q_3\|\psi_1 - \psi_2\|_C. \tag{5.39}$$

The operator Q is thus a contraction in $\omega_1(r, \rho, \gamma)$, and equation (5.32) in $\omega_1(r, \rho, \gamma)$ has a unique solution.

Suppose now that the point $x_0 = (x_1^0, x_2^0) \in Y_1$, i.e., $x_2^0 = \psi(x_1^0)$. We shall show that then there exists a point $x_{-1} = (x_1^{-1}, x_2^{-1})$ such that $N x_{-1} = x_0$. According to (5.2), this can be rewritten in the form

$$\begin{aligned} U_1 x_1^{-1} + f_1(x_1^{-1}, x_2^{-1}) &= x_1^0; \\ U_2 x_2^{-1} + f_2(x_1^{-1}, x_2^{-1}) &= x_2^0. \end{aligned} \tag{5.40}$$

If system (5.40) has a solution, then by the invariance of the manifold Y_1 the relation $x_2^{-1} = \psi(x_1^{-1})$ must hold. The first equation of (5.40) then takes the form

$$U_1 x_1^{-1} + f_1(x_1^{-1}, \psi(x_1^{-1})) = x_1^0. \tag{5.41}$$

Conversely, if (5.41) is solvable, then we set $x_2^{-1} = \psi(x_1^{-1})$. The first equation of (5.40) is then obvious, and we prove the second by using (5.32) and (5.41):

$$\begin{aligned} U_2 \psi(x_1^{-1}) + f_2(x_1^{-1}, \psi(x_1^{-1})) &= \psi(U_1 x_1^{-1} + f_1(x_1^{-1}, \psi(x_1^{-1}))) \\ &= \psi(x_1^0) = x_2^0. \end{aligned}$$

It thus suffices to consider (5.41). We reduce it to the form

$$x_1^{-1} = U_1^{-1} x_1^0 - U_1^{-1} f_1(x_1^{-1}, \psi(x_1^{-1})). \tag{5.42}$$

This equation essentially coincides with (5.29) (in the latter it suffices to make the change $x_1 \to x_1^{-1}$, $z \to x_1^0$). As we proved earlier, it is uniquely solvable. From (5.30) we obtain

$$\|x_1^{-1}\| \le q\|x_1^0\|; \qquad (q_1 < q < 1). \tag{5.43}$$

Noting that $x_2^{-1} = \psi(x_1^{-1})$ and $\psi \in \omega_1(r, \rho, \gamma)$, we obtain

$$\|x_2^{-1}\| \le \gamma \|x_1^{-1}\| \le \gamma q \|x_1^0\|. \tag{5.44}$$

We now define the points $x^{-n} = (x_1^{-n}, x_2^{-n})$, $n = 1, 2, \ldots$, as solutions of the equations

$$N x^{-n} = x^{-n+1}. \tag{5.45}$$

Applying inequalities of the type (5.43), (5.44), we obtain

$$\|x_1^{-n}\| \le q^n \|x_1^0\|; \qquad \|x_2^{-n}\| \le \gamma^n q^n \|x_1^0\|. \tag{5.46}$$

We thus conclude that $x^{-n} = N^{-n} x_0 \to 0$ $(n \to \infty)$.

Suppose now that $x_0 \in D_0 - Y_1$. We assume that it is possible to construct a corresponding sequence x^{-n} (equations (5.45) are solvable). Suppose further that

$$\|x^{-n}\| \le \varepsilon_0,$$

where ε_0 is sufficiently small. We prove that then

$$\delta_n = \|x_2^{-n} - \psi(x_1^{-n})\| \ge \kappa^n \delta_0; \qquad \kappa > 1. \tag{5.47}$$

Indeed, applying (5.45) and (5.46), we obtain

$$\delta_n = \|U_2(x^{-n-1} - \psi(x_1^{-n-1})) + f_2(x_1^{-n-1}, x_2^{-n-1}) \\ - f_2(x_1^{-n-1}, \psi(x_1^{-n-1}))\|. \tag{5.48}$$

Further, we have

$$\delta_n \le [q_2 + m_{22}(\varepsilon_0, \varepsilon_0)] \delta_{n+1}. \tag{5.49}$$

Estimate (5.47) follows directly from (5.49). Here as κ it is possible to take any number less than $1/q_2$ and on the basis of it to choose ε_0 so that (5.47) holds.

Thus, if x_0 does not belong to the manifold Y_1, then either the points $N^{-n} x_0$, beginning with some n, are not defined, or they leave some ε_0-neighborhood of zero of the space X (ε_0 is completely determined by the operator N and does not depend on x_0).

Lemma 5.1 is thus completely proved. The next theorem is a direct corollary of it.

THEOREM 5.1. *Suppose the stability spectrum of the periodic flow $v_0(t)$ can be represented as the union of nonintersecting closed sets Σ_1 and Σ_2, the first of which is situated in the left half-plane and the second in the right half-plane. In a neighborhood of zero of the space H_1 a finite-dimensional manifold Y_1 and a manifold of finite codimension Y_2 are then defined which possess the following properties:*

1) *If $a \in Y_2$, then $N^t a \to 0$ in H_1 as $t \to +\infty$.*

2) *If the initial vector $a \in H_1$ has sufficiently small norm and $a \notin Y_2$, then $N^t a$ in time leaves some fixed neighborhood of zero.*

3) *If $a \in Y_1$, then the Cauchy problem* (1.1), (2.14) *has a solution defined for all $t < 0$, and $N^t a \to 0$ in H_1 as $t \to -\infty$.*

4) *If $a \in H_1$ has small norm and does not lie on the manifold Y_1, then $N^{-t} a$ for sufficiently large t leaves some fixed neighborhood of zero of the space H_1 (in particular, this can mean that the solution of the Cauchy problem is not defined form $t < t_0 \leq 0$ for some t_0).*

5) *The tangent manifolds to Y_1 and Y_2 are invariant subspaces of the monodromy operator corresponding to the components of its spectrum situated, respectively, outside and inside the unit disk.*

We make some remarks on Theorems 3.1, 4.1, and 5.1.

1. If there are multipliers on the unit circle, the previous method makes it possible to prove the existence of an invariant manifold Y_2 (Y_1) on which stability is preserved as $t \to +\infty$ ($t \to -\infty$) and which is tangent to the invariant subspace of the monodromy operator corresponding to the part of the spectrum within (outside) the unit disk. However, it remains unknown how trajectories not lying on Y_1 or Y_2 behave—here an additional investigation is required with consideration of the character of the nonlinear terms. The works of Kelley [31] and Hirsch, Pugh, and Shub [23] are devoted to invariant manifolds in the critical case; see also [50].

2. The choice of metric of H_1 was occasioned only by our desire to make the derivation elementary. All results carry over to any space of initial data for which (local) solvability of the Cauchy problem for the Navier-Stokes equations and continuous dependence on the initial data can be proved.

3. It would not be hard to investigate the smoothness of the manifolds Y_1 and Y_2 and prove that the operators φ and ψ defined in (5.3) and (5.25) have Fréchet derivatives of all orders.

4. The stability or instability of the periodic regime v_0 under the conditions of Theorems 3.1–5.1 does not depend on the choice of the initial time. Suppose, for example, that perturbations occur at time t_0. We denote the corresponding translation operator along trajectories by

$U_{t_0,t}$: $U_{t_0,t}u(t_0) = u(t_0 + t)$. The new monodromy operator is $U_{t_0,T}$. We shall assume that $0 \leq t_0 < T$—this can always be arranged by going over from t_0 to $t_0 + kT$, where k is an integer; here $U_{t_0+T,T} = U_{t_0,T}$. We shall show that the spectrum of the monodromy operator does not depend on t_0. This follows from the obvious equalities

$$U_{t_0,T} = U_{t_0} U_{t_0,T-t_0}; \qquad U_T = U_{t_0,T-t_0} U_{t_0} \tag{5.50}$$

on the basis of the following lemma (the first assertion of it is a known fact of the theory of normed rings; see, for example, [108]).

LEMMA 5.2. *Let $U, V : X \to X$ be bounded linear operators acting in the Banach space X. The operators UV and VU then have the same nonzero spectrum:*(15)

$$\sigma(UV) = \{0\} = \sigma(VU) - \{0\}. \tag{5.51}$$

Moreover, if 0 is excluded, then the point, continuous, and residual spectra, respectively, coincide:

$$\sigma_p(UV) - \{0\} = \sigma_p(VU) - \{0\}; \qquad \sigma_c(UV) - \{0\} = \sigma_c(VU) - \{0\};$$
$$\sigma_r(UV) - \{0\} = \sigma_r(VU) - \{0\}. \tag{5.52}$$

Moreover, the multiplicities of nonzero eigenvalues coincide.

PROOF. Suppose $\lambda \neq 0$ and $\lambda \in \sigma_p(UV)$. This means that there exists an element $x_0 \in X$ such that $(\lambda I - UV)x_0 = 0$. Then, obviously, $y_0 = Vx_0 \neq 0$ and $(\lambda I - VU)y_0 = 0$. Thus, $\sigma_p(UV) - \{0\} \subset \sigma_p(VU) - \{0\}$. The opposite inclusion can be obtained similarly, and the first relation of (5.2) is proved. If x_1 is an associated vector of the operator $UV : (\lambda I - UV)x_1 = x_0$ (x_0 is an eigenvector and λ is a nonzero eigenvalue), then for $y_1 = Vx_1$ we have $(\lambda I - VU)y_1 = Vx_0 = y$. Therefore, y_1 is an associated vector of the operator VU. The one-to-one correspondence of higher associated vectors can be established in exactly the same way. Thus, the multiplicity of the eigenvalue λ for the operators UV and VU coincide.

Applying familiar relations between the spectra of an operator and its adjoint ([15], Chapter VII, §5, Exercise 9) and the part of the lemma already proved, we obtain

$$\sigma_r(UV) - \{0\} \subset \sigma_p(V^*U^*) - \{0\} = \sigma_p(U^*V^*) - \{0\} \subset \sigma_r(VU) - \{0\}. \tag{5.53}$$

The third relation of (5.52) follows from this.

---

(15)The point 0 can belong to $\sigma(UV)$ and not belong to $\sigma(VU)$. Example: $X = L_2$, $Ux = \sum_1^\infty \xi_{2k} e_k$, $Vx = \sum_1^\infty \xi_k e_{2k}$, and $x = \sum_1^\infty \xi_x e_k$, where $e_1, e_2, \ldots$ is an orthonormal basis. Then $UV = I$, but $VUx = \sum_1^\infty \xi_{2k} e_{2k}$ is not invertible.

Finally, suppose $\lambda \in \sigma_c(UV) - \{0\}$. It follows that there exists a sequence $x_n \in X$ such that $\|x_n\| = 1$ and $\|(\lambda I - UV)x_n\| \to 0$ $(n \to \infty)$. It is clear that $\|Vx_n\| \geq \alpha > 0$ for all n beginning with some one; α does not depend on n. Setting $y_n = Vx_n/Vx_n\|$, we find that $\|y_n\| = 1$ and

$$\|(\lambda I - VU)y_n\| \leq (1/\alpha)\|V\| \cdot \|(\lambda I - UV)x_n\| \to 0.$$

Thus, $\sigma_c(UV) - \{0\} \subset \sigma_c(VU) - \{0\}$. Applying the part of the lemma already proved, we arrive at the second equality of (5.52). Lemma 5.2 is proved.

Since 0 clearly belongs to the spectrum of each monodromy operator U_{T,t_0} in view of its compactness, from Lemma 5.2 and (5.50) it actually follows that $\sigma(U_{T,t_0})$ does not depend on t. This can, by the way, be proved without using compactness. Indeed, the equality

$$U_{T,t_0}^{-1} = U_{t_0} U_T^{-2} U_{T-t_0,T+t_0}$$

shows that the monodromy operators U_{T,t_0} are either invertible for all t_0 or noninvertible for all t_0.

It can be proved that the manifolds Y_1 and Y_2 vary smoothly with changing t_0 if the data of the problem are sufficiently smooth.

§6. Stability of self-oscillatory regimes

We consider the Navier-Stokes equations in a bounded three-dimensional domain Ω with boundary S of class C^2:

$$\frac{\partial v}{\partial t} + (v, \nabla)v - \Delta v = -\nabla P + F(x); \tag{6.1}$$

$$\operatorname{div} v = 0; \tag{6.2}$$

$$v/S = \alpha(x). \tag{6.3}$$

We assume that the velocity α on the boundary of the domain and the mass forces do not depend on time. We suppose there exists a solution (T-periodic in time) $v_0(x,t)$, $P_0(x,t)$ of system (6.1), (6.2)—an auto-oscillation. Then $v_0(x, t+h)$, $P_0(x, t+h)$ is also a T-periodic solution for any h. Clearly asymptotic stability is here impossible: for example, by taking at the initial time the vector v equal to $v_0(x,h)$, we land on the neighboring periodic regime. It is therefore natural to consider stability of an auto-oscillatory solution as an invariant set—a cycle $\bigcup_h v_0(t+h) = C$.

DEFINITION. We say that a cycle C is *stable in* H_1 if for any $\varepsilon > 0$ it is possible to find $\delta > 0$ such that the solution of the Cauchy problem (1.1), (2.14) for all $t > 0$ satisfies

$$\rho(v(t), C) \equiv \inf_{0 \leq S \leq T} \|v_0(t) + u(t) - v_0(s)\|_{H_1} < \varepsilon, \tag{6.4}$$

provided that

$$\rho(v(0), C) \equiv \inf_{0 \leq S \leq T} \|v_0(0) + a - v_0(s)\|_{H_1} < \delta. \tag{6.5}$$

We say that C is *asymptotically stable* if, additionally, as $t \to \infty$

$$\rho(v(t), C) \to 0. \tag{6.6}$$

We suppose that $v_0 \in \widetilde{H}_2^T$ and $dv_0/dt \in H_2^T$. The spectral problem (1.5) then admits the solution $\sigma_0 = 0$, $w_0(t) = dv_0(t)/dt$. Correspondingly, the monodromy operator U_T has the eigenvalue $\rho_0 = 1$, to which there corresponds the eigenvector $\varphi = w_0(0)$.

THEOREM 6.1. *Suppose* 1 *is a simple eigenvalue of the monodromy operator U_T while all the remaining points of the spectrum $\sigma(U_T)$ lie inside the unit disk. Then the cycle C is exponentially asymptotically stable. Moreover, if $\rho(v(0), C)$ is sufficiently small, then there exists a number $h_0 = h(v(0))$ (the asymptotic phase) such that*

$$\|v(t) - v_0(t + h_0)\|_{H_1} \to 0, \qquad (t \to +\infty). \tag{6.7}$$

The proof follows from the lemma presented below.

Suppose there is given a dynamical system $(\widetilde{N}_t, X)$, i.e., a nonlinear partial semigroup of operators $\widetilde{N}_t : X \to X$ $(0 \leq t < \infty)$ acting in a Banach space X. This means that for each $x \in X$ the element $\widetilde{N}_t x$ is defined for $0 < t < t_0(x)$, and $\widetilde{N}_0 = I$ and $\widetilde{N}_{t+\tau} x = \widetilde{N}_t \widetilde{N}_\tau x$ $(t, \tau \geq 0, t + \tau < t_0(x))$. We suppose that $\widetilde{N}_T$ for some $T > 0$ has a fixed point q_0:

$$\widetilde{N}_T q_0 = q_0. \tag{6.8}$$

Then $q_\tau = \widetilde{N}_\tau q_0$ is also a fixed point of $\widetilde{N}_T$ for all $\tau > 0$. Indeed, it follows from condition (6.8) that q_τ is defined for all $\tau \in [0, \infty]$. Further, $\widetilde{N}_T(\widetilde{N}_\tau q_0) = \widetilde{N}_\tau \widetilde{N}_T q_0 = \widetilde{N}_\tau q_0$.

We suppose that for all $\tau \geq 0$

$$\widetilde{N}_{T+\tau} q_0 = \widetilde{N}_\tau q_0. \tag{6.9}$$

If $\bigcup_{0 \leq \tau \leq T} q_\tau = C$ contains more than one point, we say that C is a *cycle* (or an *auto-oscillation*) of the dynamical system $(\widetilde{N}_t, X)$. If $C = \{q_0\}$, then q_0 is called a *fixed point* (or an *equilibrium*) of the dynamical system $(\widetilde{N}_t, X)$.

We say that a cycle $C = \bigcup_{0 \leq \tau \leq T} \widetilde{N}_\tau q_0$ of the dynamical system $(\widetilde{N}_t, X)$ is *stable* if for any $\varepsilon > 0$ it is possible to find a $\delta > 0$ such that $\rho(x_0, C) < \delta$ implies that the point $\widetilde{N}_t x_0$ is defined for all $t > 0$ and $\rho(\widetilde{N}_t x_0, C) < \varepsilon$ for $t > 0$. We say that a cycle C is *asymptotically stable* if, additionally, $\rho(\widetilde{N}_t x_0, C) \to 0$ as $t \to \infty$. If, moreover, there exists $\tau_0 = \tau_0(x_0)$ such that

$\rho(\widetilde{N}_t x_0, \widetilde{N}_t + \tau_0 q_0) \to 0$ as $t \to \infty$, then we say that the trajectory $\widetilde{N}_t x_0$ has an *asymptotic phase*.

Further, we call a cycle C *smooth* if in some neighborhood $\omega\delta = \{x \in X : \rho(x, C) < \delta\}$ of it the operators $\widetilde{N}_t$ $(0 \le t \le T)$ are continuously Fréchet differentiable, the derivative $dq_\tau/d\tau = \dot{q}_\tau$ $(0 \le t \le T)$ exists, and, moreover, for any $\varepsilon_1, \varepsilon_2 > 0$ there exist $\delta_1, \delta_2 > 0$ such that $\|a\|_x < \delta_1$ and $|S| < \delta_2$ imply the following estimates which are uniform with respect to $\tau \in [0, T]$:

$$\|\Delta_1(\tau, a)\| = \|\widetilde{N}_T(q_\tau + a) - q_\tau - \widetilde{N}'_T(q_\tau)a\| \le \varepsilon_1\|a\|; \tag{6.10}$$

$$\|\Delta_2(\tau, S)\| = \|q_{\tau+S} - q_\tau - \dot{q}_\tau S\| \le \varepsilon_2|S|. \tag{6.11}$$

We remark that $\dot{q}_t \neq 0$ for any t. Indeed, suppose $q_{t_0} = 0$. We shall show that then $\widetilde{N}_t q_{t_0} = q_{t_0}$ for all $t > 0$. Indeed, for any $\varepsilon_2 > 0$ we choose a number $\delta_2 > 0$ such that (6.11) holds. Suppose the natural number n is so large that $t/n < \delta_2$. We have

$$\begin{aligned} \|q_{t_0+t} - q_{t_0}\| &\le \sum_{k=1}^{n} \left\| q_{t_0+\frac{kt}{n}} - q_{t_0+\frac{(k-1)t}{n}} \right\| \\ &\le \varepsilon_2 \frac{t}{n} \cdot n = \varepsilon_2 t, \end{aligned}$$

whence it follows that $q_{t_0+t} = q_{t_0}$, since ε_2 is arbitrarily small and does not depend on t.

We introduce the notation $U_{t,\tau} = \widetilde{N}'_t(q_\tau)$. We call the operator $U_{T,\tau}$ the *monodromy operator* (corresponding to the initial time τ). It is easy to see that $\varphi_\tau = \dot{q}_\tau$ is an eigenvector of $U_{T,\tau}$ corresponding to the eigenvalue 1:

$$U_{T,\tau}\varphi_\tau = \varphi_\tau. \tag{6.12}$$

To prove (6.12) it suffices to differentiate (6.9) with respect to τ. Differentiating the equality $q_\tau = \widetilde{N}_\tau q_0$ with respect to τ, we obtain

$$\varphi_\tau = \widetilde{N}'_\tau(q_0)\varphi_0. \tag{6.13}$$

In general if φ_0 is some eigenvector of $U_{T,0}$, then the vector φ_τ defined by (6.13) is an eigenvector of $U_{T,\tau}$. This follows directly from the identity

$$\widetilde{N}'_{s+\tau}(q_0) = \widetilde{N}'_s(q_\tau)\widetilde{N}'_\tau(q_0) = \widetilde{N}'_\tau(q_s) \cdot \widetilde{N}'_s(q_0); \quad (s, \tau \ge 0), \tag{6.14}$$

which we obtain by taking the Fréchet derivative of the equality $\widetilde{N}_{s+\tau}x = \widetilde{N}_s\widetilde{N}_\tau x = \widetilde{N}_\tau\widetilde{N}_s x$ at the point $x = q_0$; in this identity it is necessary to set $s = 0$ and note that $\widetilde{N}_T q_0 = q_0$.

Further, suppose that φ_0 and ψ_0 are eigenvectors of the operators $U_{T,0}$ and $U^*_{T,0}$ to which there correspond eigenvalues ρ_1, and ρ^*_2 respectively. We define φ_τ by (6.13) and ψ_τ by the equality

$$\psi_\tau = \frac{1}{\rho^*_2}\widetilde{N}'^*_{T-\tau}(\widetilde{N}_\tau q_0)\psi_0; \qquad (0 \le \tau \le T). \tag{6.15}$$

Then ψ_τ is an eigenvector of $U^*_{T,\tau}$, and $(\varphi_\tau, \psi_\tau)$ does not depend on τ. Indeed, applying (6.14) for $s = T - \tau$, we obtain

$$\begin{aligned} U^*_{T,\tau}\psi_\tau &= \frac{1}{\rho^*_2}\widetilde{N}'_{T^*}(q_\tau)\widetilde{N}'_{T^*-\tau}(\widetilde{N}_\tau q_0) \\ &= \frac{1}{\rho^*_2}\widetilde{N}'_{T^*-\tau}(q_\tau)\widetilde{N}'_{T^*}(q_0) = \rho^*_2\psi_\tau. \end{aligned} \tag{6.16}$$

Here $(\varphi_\tau, \psi_\tau)$ does not depend on τ. Indeed, again applying (6.14), we obtain

$$\begin{aligned} (\varphi_\tau, \psi_\tau) &= \frac{1}{\rho_2}(\widetilde{N}'_{T-\tau}(q_\tau)\widetilde{N}'_\tau(q_0)\varphi_0, \psi_0) \\ &= \frac{1}{\rho_2}(\widetilde{N}'_T(q_0)\varphi_0, \psi_0) \\ &= \frac{\rho_i}{\rho_2}(\varphi_0, \psi_0). \end{aligned} \tag{6.17}$$

If $\rho_1 \neq \rho_2$, then the right side of (6.17) vanishes.

An isolated eigenvalue λ_0 of a linear operator $U : X \to X$ is called *simple* if the eigensubspace corresponding to it is one-dimensional and λ_0 is a simple pole of the resolvent of U. In this case λ^*_0 is a simple eigenvalue of the adjoint operator, and the corresponding eigenvectors are not orthogonal.

We remark that the spectrum of the monodromy operator $U_{T,\tau}$ does not depend on τ (see Lemma 5.2 and the remarks to it). The multiplicity of the eigenvalue 1 is the same for all τ.

LEMMA 6.1. *Let $C = \bigcup_{0\le\tau\le T}\widetilde{N}_\tau q_0$ be a smooth cycle of the dynamical system $(\widetilde{N}_t, X)$, and suppose the spectrum of the monodromy operator $U_{T,\tau}$ $(0 \le \tau \le T)$ has the form $\sigma(U_{T,\tau}) = \{1\} \cup \sigma_0(U_{T,\tau})$, where 1 is a simple isolated eigenvalue and*

$$|\sigma_0(U_{T,\tau})| < \alpha < 1. \tag{6.18}$$

Then the cycle C is asymptotically stable, and any trajectory $\{\widetilde{N}_t x_0\}$ $(t \ge 0)$ has asymptotic phase provided that $\rho(x_0, C)$ is sufficiently small.

PROOF. Suppose $\varphi_0 = \dot{q}_0$ and ψ_0 are fixed vectors of the respective operators $U_{T,0}$ and $U^*_{T,0}$ with $(\varphi_0, \psi_0) = 1$. We define φ_τ and ψ_τ by (6.13) and (6.15) for $\rho = 1$. Then $\varphi_\tau = \dot{q}_\tau$ and ψ_τ are eigenvectors of the operators $U_{T,\tau}$ and $U^*_{T,\tau}$ with eigenvalue 1. Moreover, by (6.17) we have $(\varphi_\tau, \psi_\tau) = 1$ $(0 \le \tau \le T)$. We define V_τ by setting

$$V_\tau x = U_{T,\tau} x - (x, \psi_\tau)\varphi_\tau. \tag{6.19}$$

then $\sigma(V_\tau) = \sigma_0(U_{T,\tau}) \cup \{0\}$, and by (6.18) we have

$$|\sigma(V_\tau)| < \alpha < 1. \tag{6.20}$$

We show that there exists a natural number m such that the operators V_τ^m for all $0 \le \tau \le T$ are contractions, and, moreover, the uniform estimate

$$\|V_\tau^m\| \le \Theta < 1. \tag{6.21}$$

holds. Indeed, from the smoothness of the cycle C it follows that the operator-valued function V_τ is continuous in τ in the uniform operator topology. Considering (6.20), we conclude from this that the resolvent $(\lambda I - V_\tau)^{-1}$ is jointly continuous in (λ, τ): $|\lambda| = \alpha$, $0 \le \tau \le T$ (see [15], Chapter VII, §6). We thus have

$$\max_{|\lambda|=\alpha} \max_{0 \le \tau \le T} \|(\lambda I - V_\tau)^{-1}\| = h < \infty. \tag{6.22}$$

We now use the representation of V_τ^m in the form

$$V_\tau^m = \frac{1}{2\pi i} \int_{|\lambda|=\alpha} \lambda^m (\lambda I - V_\tau)^{-1} \, d\lambda. \tag{6.23}$$

From (6.22) and (6.23) we conclude that $\|V_\tau^m\| \le h\alpha^m \to 0$ $(m \to \infty)$. Thus, if $0 < \Theta < 1$ it is possible to choose m so that (6.21) holds. It may be assumed with no loss of generality that $m = 1$; we can go over to this case by the change $T \to mT$, which has no effect on what follows.

We thus assume that

$$\|V_\tau\| < \Theta < 1; \qquad (0 \le \tau \le T). \tag{6.24}$$

Suppose x_0 is a point in a δ-neighborhood of the cycle C: $\rho(x_0, C) < \delta$. Then for some τ_0 we have

$$\|x_0 - q_{\tau_0}\| < \delta; \qquad (0 \le \tau_0 \le T). \tag{6.25}$$

We now set $x_n = \widetilde{N}_T^n x_0$ $(n = 0, 1, \dots)$, and we estimate $\rho(x_n, C)$. For this we consider the sequence $\{\tau_n\}$ of times and the sequence $\{a_n\}$ of elements of the space X defined by

$$\tau_{n+1} = \tau_n + s_n; \quad s_n = (a_n, \psi_{\tau_n}); \quad a_n = x_n - q_{\tau_n}; \qquad (n = 0, 1, 2, \dots). \tag{6.26}$$

We show that if δ is sufficiently small then

$$\rho(x_n, C) \leq \|a_n\| \leq \Theta^n \delta; \qquad |\tau_{n+1} - \tau_n| \leq l\|a_n\| \leq l\delta\Theta^n, \tag{6.27}$$

where

$$l = \max_{0 \leq \tau \leq T} \|\psi_\tau\|x^*.$$

Suppose $\delta_1, \delta_2 > 0$ are such that (6.10) and (6.11) hold. We consider $\varepsilon_1, \varepsilon_2$, and δ so small that

$$\max_{0 \leq \tau \leq T} \|V_\tau\| + \varepsilon_1 + \varepsilon_2 l < \Theta; \tag{6.28}$$
$$\delta < \delta_1; \qquad l\delta < \delta_2.$$

We apply induction on n. For $n = 0$ the first estimate of (6.27) is satisfied, and the second follows from the first:

$$|\tau_1 - \tau_0| = |(a_0, \psi_{\tau_0})| \leq l\delta. \tag{6.29}$$

We now prove the $(n+1)$st estimate of (6.27) under the assumption that the nth has already been proved. We have

$$\rho(x_{n+1}, C) \leq \|a_{n+1}\| = \|\widetilde{N}_T(q_{\tau_n} + a_n) - q_{\tau_n + s_n}\|. \tag{6.30}$$

We next use (6.10) and (6.11). The conditions for their applicability are satisfied: by (6.29)

$$\|a_n\| < \delta < \delta_1; \qquad |s_n| = |(a_n, \psi_{\tau_n})| \leq \|a_n\| \cdot \|\psi_{\tau_n}\| \leq l\delta < \delta_2. \tag{6.31}$$

Thus, from (6.30), also considering (6.28), we conclude that

$$\begin{aligned}\rho(x_{n+1}, C) &\leq \|a_{n+1}\| \\ &= \|V_{\tau_n} a_n + \Delta_1(\tau_n, a_n) + \Delta_2(\tau_n, s_n)\| \\ &\leq \Theta\|a_n\|.\end{aligned} \tag{6.32}$$

Applying (6.32), we now obtain

$$|\tau_{n+1} - \tau_n| = |(a_n, \psi_n)| \leq l\|a_n\| \leq l\delta\Theta^n. \tag{6.33}$$

From (6.32) and (6.33) we obtain the $(n+1)$st estimate of (6.27) and thus complete the proof of it for any n.

From the first inequality of (6.27) it follows that the points x_n converge to the cycle C, while from the second it follows that the sequence τ_n converges:

$$h = \tau_0 + \sum_{n=0}^{\infty} (\tau_{n+1} - \tau_n) = \lim_{n \to \infty} \tau_n. \tag{6.34}$$

Setting $\tau_n = h + \eta_n$, from (6.27) we conclude that

$$\tau_n = h + \eta_n; \qquad |\eta_n| \leq \frac{l\delta}{1-\Theta}\Theta^n. \tag{6.35}$$

In order to complete the proof of asymptotic stability of the cycle C and the existence of an asymptotic phase, it suffices to establish that

$$\|\widetilde{N}_t x_0 - q_{h+t}\| \to 0, \qquad (t \to +\infty). \tag{6.36}$$

Let $t = nT + s$, $0 \leq s < T$. In the notation of (6.10) and (6.11) the left side of (6.36) then takes the form

$$\begin{aligned} \|\widetilde{N}_t x_0 - q_{h+t}\| &= \|\widetilde{N}_s x_n - q_{h+s}\| \\ &= \|\Delta_1(\tau_n, a_n) + \widetilde{N}_s'(q_{\tau_n})a_n + \Delta_2(h+s, \eta_n) + \dot{q}_{h+s}\eta_n\|. \end{aligned} \tag{6.37}$$

Estimating this expression by means of (6.10), (6.11), (6.33), and (6.35), we obtain

$$\|\widetilde{N}_t x_0 - q_{h+t}\| \leq m\delta\Theta^n, \tag{6.38}$$

where m is a constant no depending on n and defined by

$$\begin{gathered} m = m_1 + \varepsilon_1 + \frac{m_2 + \varepsilon_2}{1-\Theta}; \\ m_1 = \sup_{0 \leq s, \tau \leq T} \|\widetilde{N}_s'(q_\tau)\|; \qquad m_2 = \sup_{0 \leq \tau \leq T} \|\dot{q}_\tau\|. \end{gathered} \tag{6.39}$$

The quantities m_1 and m_2 are finite in view of the smoothness of C. From (6.38) we obtain (6.36). Lemma 6.1 is proved.

We further note an estimate of the rate of approach to the limiting periodic regime which follows from (6.38):

$$\|\widetilde{N}_t x_0 - q_{h+t}\| \leq m\delta e^{-\sigma_0 t}; \qquad \sigma_0 = \frac{l}{T}\ln\frac{1}{\Theta}. \tag{6.40}$$

Theorem 6.1 is easily derived from Lemma 6.1. The smoothness of the cycle follows from Lemmas 2.4 and 2.5. The operators $\widetilde{N}_t$ are connected with the operators N_t by the equality

$$N_t a = \widetilde{N}_t(v_0(0) + a) - v_0(t); \qquad a \in H_1. \tag{6.41}$$

In the case where the boundary condition (6.3) is inhomogeneous it suffices to consider the semigroup $\widetilde{N}_t$ to be defined only on vectors of the form $v_0(0) + a$, $a \in H_1$; in the foregoing considerations nothing changes if in place of the Banach space X we take the "hyperspace" X_0—the set of elements of some Banach space defined by the condition $Qx = \alpha$, where Q is a given linear operator and α is a given element. Of course, it is still more natural to consider the semigroup $\widetilde{N}_t$ on a Banach manifold. Moreover, using the change $v = v' + \alpha'$ (α' is a smooth vector satisfying condition (6.3)), it is sufficient in our case to make the condition on the boundary homogeneous in order to obtain precisely the situation of Lemma 6.1.

We further formulate the condition of Theorem 6.1 in terms of the stability spectrum of the periodic motion $v_0(t)$: equation (1.3) and the adjoint equation

$$\frac{d\tilde{u}}{dt} - A_0\tilde{u} - B^*(t)\tilde{u} = 0 \tag{6.42}$$

each have precisely one (up to a constant factor) T-periodic solution $u_0(t)$, $\tilde{\mathbf{u}}_0(t)$, which are not orthogonal:

$$\int_0^T (u_0(t), \tilde{\mathbf{u}}_0(t))_H \, dt \neq 0. \tag{6.43}$$

The remaining points of the stability spectrum (aside from $\sigma_0 = 0$) must be located within the left half-plane.

§7. Instability of cycles

Theorem 4.1 contains sufficient conditions for the instability of an individual periodic regime. It turns out that under the same conditions as in the case of auto-oscillations it is possible to assert something more—instability of a cycle.

THEOREM 7.1. *If the stability spectrum of an auto-oscillatory T-periodic regime $v_0(t)$ contains at least one eigenvalue σ with positive real part (or, equivalently, there exists a multiplier ρ with $|\rho| > 1$), then the cycle $C = \bigcup_{0\le\tau\le T} v_0(dt)$ is unstable in H_1.*

We shall derive this theorem from a more general assertion.

LEMMA 7.1. *Let $C = \bigcup_{0\le\tau\le T} \widetilde{N}_\tau q_0$ be a smooth cycle of the dynamical system $(\widetilde{N}_t, X)$, and suppose that the spectrum of the monodromy operator $U_{T,\tau}$ can be represented as the union of nonintersecting closed sets $\sigma_1(U_{T,\tau})$ and $\sigma_2(U_{T,\tau})$, where*

$$|\sigma_1(U_{T,\tau})| > \beta > 1; \qquad |\sigma_2(U_{T,\tau})| \le 1. \tag{7.1}$$

Suppose the mapping $\widetilde{N}_t$ is differentiable with respect to t and the derivative $\dot{\widetilde{N}}_t$ is continuous in $(x,t) : |t-T| < \delta_1$ and $\rho(x, C) < \delta_2$ for some $\delta_1, \delta_2 > 0$. Then the cycle C is unstable.

PROOF. We break the proof into several steps.

I. *Suppose T is the least period; then the mapping $t \to q_t$ of the circle (realized as the segment $[0, T]$ with the end points identified) into X has* *a uniformly continuous inverse: for any $\delta > 0$ there exists $\varepsilon > 0$ such that $\|q_t - q_\tau\| < \delta$ $(0 \le t \le \tau \le T)$ implies that either $|t-\tau| < \delta$ or $|t-\tau+T| < \delta$.*

If this were not true, then, using the compactness of the circle, it would be possible to find convergent sequences $t_n \to t_0$ and $\tau_n \to \tau_0$ $(0 \le t_n \le \tau_n \le T)$ and a number $\varepsilon_0 > 0$ such that either $|t_n - \tau_n| \ge \varepsilon_0$ or $|T + t_n - \tau_n| \ge \varepsilon_0$, and $q_{t_n} - q_{\tau_n} \to 0$. But then $q_{t_0} = q_{\tau_0}$ and at least one of the following two inequalities holds:

$$|t_0 - \tau_0| \ge \varepsilon_0; \qquad |T + t_0 - \tau_0| \ge \varepsilon_0. \tag{7.2}$$

Since either $t_0 = \tau_0$ or $t_0 + T = \tau_0$, it follows from (7.2) that $\varepsilon_0 \le 0$. This contradiction proves assertion I.

II. *Suppose $\psi \in X^*$ is a bounded linear functional such that*

$$(q_0, \psi) = 1, \tag{7.3}$$

Consider the set D_ε of points $x \in X$ satisfying

$$x = q_0 + a; \quad \|a\| < \varepsilon; \quad (a, \psi) = 0. \tag{7.4}$$

Then there exists a constant $\alpha > 0$ such that

$$\rho(x, C) \ge \alpha \|a\| \tag{7.5}$$

for all $x \in D_\varepsilon$.

Arguing by contradiction, we suppose that there exists a sequence $x_n \in X$ such that

$$\begin{gathered} x_n = q_0 + a_n; \quad \|a_n\| < \varepsilon; \quad (a_n, \psi) = 0; \\ \rho(x_n, C) = \|x_n - q_{t_n}\| < \frac{1}{n}\|a_n\|. \end{gathered} \tag{7.6}$$

The existence of the points q_{t_n} follows from the compactness of the cycle. Passing, if necessary, to a subsequence, we can assume that the sequence t_n converges: $t_n \to t_0$. From (7.6) it follows that $\varphi_n = q_0 + a_n - q_{t_n} \to 0$. Therefore, $a_n \to q_{t_0} - q_0$, and

$$\|q_{t_0} - q_0\| < \varepsilon; \qquad (q_{t_0} - q_0, \psi) = 0. \tag{7.7}$$

It follows from assertion I that for small ε the number t_0 is close to 0 or to T. Suppose, for example, that t_0 is small. Then $q_{t_0} - q_0 = \dot{q}_0 t_0 + o(t_0)$, and form (7.7) we find that

$$t_0 + o(t_0) = 0. \tag{7.8}$$

It follows from (7.8) that if ε is sufficiently small, then $t_0 = 0$.

Thus, possibly passing again to a subsequence, we can assume that $t_n \to 0$ (the case $t_n \to T$ can be considered in the same way). Using the smoothness of the cycle and the last relation of (7.6), we now obtain

$$\frac{1}{n}\|a_n\| \cdot \|\psi\| > |(\varphi_n, \psi)| = |(a_n - \dot{q}_0 t_n + o(t_n), \psi)| \\ = t_n + o(t_n). \tag{7.9}$$

From (7.9) it follows that there exists a constant γ such that

$$0 < t_n < \frac{\gamma}{n}\|a_n\|. \tag{7.10}$$

Again applying (7.6), we obtain

$$\frac{1}{n}\|a_n\| > \|\varphi_n\| > \|a_n\| - t_n\|\dot{q}_0\| + o(t_n). \tag{7.11}$$

The contradiction between (7.10) and (7.11) proves assertion II.

III. Arguing by contradiction, we suppose that the cycle C is stable. Then for any $\varepsilon > 0$ it is possible to find $\delta > 0$ such that $\rho(x_0, C) < \delta$ implies

$$\rho(\widetilde{N}_t x_0, C) < \alpha\varepsilon; \qquad (t \geq 0). \tag{7.12}$$

Suppose $x = q_0 + a \in D_\varepsilon$ *and* ε *is sufficiently small. Then there exists* $t_* = t_*(x) > 0$ *such that* $\widetilde{N}_{t_*} x \in D_\varepsilon$. *Here*([16])

$$t_*(x) = T - (\widetilde{N}'_T(q_0)a, \psi) + o(a). \tag{7.13}$$

This assertion means that a trajectory with an initial point in D_ε over a course of time close to the period T returns to this set.([17])

We consider the function

$$f(s) = (\widetilde{N}_{T+s}(q_0 + a) - q_0, \psi) \tag{7.14}$$

of the numerical argument s. We assume that s is subject to the condition $|s| \leq k\|a\|$ (k is a fixed, sufficiently large constant: $k > \|\widetilde{N}'_T(q_0)\|$). For sufficiently small a, using the smoothness of the cycle, we obtain

$$f(s) = s + (\widetilde{N}'_T(q_0), a, \psi) + \chi(s); \qquad |\chi(s)| \leq \eta(a)\|a\|; \\ \eta(a) \to 0, \qquad (\|a\| \to 0). \tag{7.15}$$

From (7.15) it follows immediately that at the end points of the interval

$$[-(\widetilde{N}'_T(q_0)a, \psi) + \eta\|a\|, -(\widetilde{N}'_T(q_0)a, \psi) - \eta\|a\|]$$

([16])If $\widetilde{N}'^{*}_T(q_0)\psi = \psi$, then $t_* = T + O(a)$, $a \to 0$.

([17])Actually, a trajectory with origin in D_ε, generally speaking, lands in $D_{\kappa\varepsilon}$ ($\kappa > 1$).

the function $f(s)$ assumes values of different signs and hence has a zero. Further, by (7.5) and (7.12) we have

$$\|\widetilde{N}_{t^*}x - q_0\| \leq \frac{1}{\alpha}\rho(\widetilde{N}_{t^*}x, C) \leq \varepsilon. \tag{7.16}$$

Thus $\widetilde{N}_t^* x \in D_\varepsilon$, and assertion III is proved.

IV. *Suppose X_0 is the subspace of X defined by the condition $(a, \psi) = 0$. D_ε is a neighborhood of zero in X_0*

Define the operator $K : D_\varepsilon \to X_0$ by setting for any $a \in D_\varepsilon$

$$Ka = \widetilde{N}_{t^*(q_0+a)}(q_0 + a) - q_0. \tag{7.17}$$

Then K is continuously differentiable in D_ε, and

$$K'(0)a = \widetilde{N}'_T(q_0)a - \dot{q}_0(\widetilde{N}'_T(q_0)a, \psi). \tag{7.18}$$

We consider the equation

$$F(t, a) \equiv (\widetilde{N}_t(q_0 + a) - q_0, \psi) = 0. \tag{7.19}$$

The mapping $F : (R \times X_0) \to X_0$ is continuously differentiable in a neighborhood of the point $(T, 0) \in R \times X_0$, and

$$F(T, 0) - 0; \qquad F_t(T, 0) = (\dot{q}_0, \psi) = 1. \tag{7.20}$$

According to the implicit function theorem, (7.19) can therefore be solved for t: there exists a mapping $\varphi : X_0 \to R$ which is uniquely determined for $a \in D_\varepsilon$ by the conditions

$$F(\varphi(a), a) = 0; \qquad \varphi(0) = T. \tag{7.21}$$

Moreover, the mapping φ is continuously differentiable, and

$$\varphi'(a)b = \frac{(\widetilde{N}'_{\varphi(a)}(q_0 + a)b, \psi)}{(\dot{\widetilde{N}}_{\varphi(a)}(q_0 + a), \psi)}. \tag{7.22}$$

In particular, for $a = 0$ we have

$$\varphi'(0)b = -(\widetilde{N}'_T(q_0)b, \psi). \tag{7.23}$$

Continuous differentiability of K is now obvious, since it is a superposition of continuously differentiable mappings. Differentiating (7.17), using (7.22), and setting $a = 0$, we obtain (7.18). Assertion IV is proved.

V. *Suppose U is a bounded linear operator acting in X and φ is an eigenvector of it corresponding to the eigenvalue ρ:*

$$U\varphi = \rho\varphi. \tag{7.24}$$

Suppose $\psi \in X^*$ *and* $(\psi, \varphi) = 1$. *Define the subspace* $X_0 = \{x \in X : (x, \psi) = 0\}$ *and the operator* $U_0 : X_0 \to X_0$ *by*

$$U_0 x = Ux - (Ux, \psi)\varphi. \tag{7.25}$$

Then the spectrum of U_0 *coincides with the spectrum of* U, *with the possible exception of the point* ρ.

Indeed, suppose λ is a regular value of U. Then for any $a \in X_0$

$$R_{0\lambda} a \equiv (\lambda I - U_0)^{-1} a = R_\lambda a - (R_\lambda a, \psi)\varphi$$
$$R_\lambda = (\lambda I - U)^{-1}. \tag{7.26}$$

Hence, λ is a regular value of U_0.

Suppose now that $\lambda \neq \rho$ is a regular value of U_0. We prove unique solvability of the equation

$$(\lambda I - U)x = f; \qquad f \in X. \tag{7.27}$$

Indeed, suppose $x_0 \in X_0$ is a solution of the equation

$$(\lambda I - U_0)x_0 = f_0; \qquad f_0 = Pf \equiv f - (f, \psi)\varphi. \tag{7.28}$$

Then the unique solution of (7.27) is

$$x = R_{0\lambda} f_0 + c\varphi; \qquad c = \frac{1}{\lambda - \rho}(f + Ux_0, \psi)\varphi. \tag{7.29}$$

We note that the resolvent R_λ has the form

$$R_\lambda f = R_{0\lambda} P f + \frac{1}{\lambda - \rho}(f + U R_{0\lambda} P f, \psi)\varphi. \tag{7.30}$$

VI. It is now not hard to finish the proof of Lemma 7.1. Indeed, from IV and V it follows that the operator K satisfies the conditions of Lemma 4.1: it is continuously differentiable in a neighborhood of zero, and the spectrum of its Fréchet differential $K'(0)$ contains the spectral set $\sigma_1(U_{T,\tau})$, with $|\sigma_1(U_{T,\tau})| > \beta > 1$.

According to Lemma 4.1, there exists $\varepsilon_0 > 0$ such that for any $\delta > 0$ it is possible to find a point $a_0 \in X_0$ and a natural number n for which

$$\|a_0\| < \delta; \qquad \|K^n a_0\| \geq \frac{1}{\alpha}\varepsilon_0. \tag{7.31}$$

but then for $x_0 = q_0 + a_0$ we have

$$\rho(x_0, C) \leq \|a_0\| < \delta; \tag{7.32}$$
$$\rho(\widetilde{N}_{t_n} x_0, C) \geq \alpha \|\widetilde{N}_{t_n} x_0 - q_0\| = \alpha \|K^n a_0\| \geq \varepsilon_0.$$

Here the number t_n is defined by

$$t_k = t_*(x_{k-1}); \qquad x_k = \widetilde{N}_{t^*(x_{k-1})} x_k \quad (k = 1, 2, \ldots, n). \tag{7.33}$$

Lemma 7.1 is thus proved.

We now say a few words regarding conditional stability of a cycle. Suppose 1 is a simple eigenvalue of the operator U. Then for the operator U_0 defined by (7.25), 1 is a regular point.

Indeed, in this case there exists a fixed vector $\varphi_* \in X^*$ of the adjoint operator U^* such that $(\varphi, \varphi_*) = 1$, while the equation

$$Ux - x = z \tag{7.34}$$

is solvable if and only if $(z, \varphi_*) = 0$ (see [15]). Suppose $y \in X_0$. Then the equation

$$U_0 x - x = y \tag{7.35}$$

has a unique solution $x \in X_0$:

$$x = x_0 + \beta\varphi; \qquad \beta = -(x_0, \psi), \tag{7.36}$$

where x_0 is some solution of (7.34) for $z = y - (y, \varphi_*)\varphi$. The operator U_0 is thus invertible on X_0.

In addition to the conditions of Lemma 7.1 we assume that the spectrum of the monodromy operator $U_{T,\tau}$ can be represented in the form of a union of the spectral sets $\sigma_1(U_{T,\tau})$ and $\sigma_2(U_{T,\tau})$, $\{1\}$, where 1 is a simple eigenvalue, and

$$|\sigma_1(U_{T,\tau})| > 1; \qquad |\sigma_2(U_{T,\tau})| < 1. \tag{7.37}$$

Suppose $\psi \in X^*$ is an eigenvector of the operator $U^*_{T,0}$ normalized by the condition $(\dot{q}_0, \psi) = 1$. We define the subspace $X_{0,\tau} \subset X$ by setting $X_{0,\tau} = \{a \in X : (a, \psi_\tau) = 0\}$, where $\psi_\tau = \widetilde{N}'^*_{T-\tau}(q_\tau)\psi = U^*_{T-\tau,\tau}\psi$ is a fixed vector of the operator $U^*_{T,\tau}$. We define the operator $U_0 : X_{0,\tau} \to X_{0,\tau}$ by

$$U_0 a = U_{T,\tau}a - (U_{T,\tau}a, \psi_\tau)\varphi; \qquad a \in X_{0,\tau}. \tag{7.38}$$

From assertion V it follows that

$$\sigma(U_0) = \sigma_1(U_{T,\tau}) \cup \sigma_2(U_{T,\tau}). \tag{7.39}$$

Suppose that $X_{1,\tau}$ and $X_{2,\tau}$ are invariant subspaces of the operator U_0 corresponding to the spectral sets σ_1 and σ_2.

Applying Lemma 5.1 to the operator K defined by (7.17), we arrive at the following proposition.

LEMMA 7.2. *Suppose that, in addition to the conditions of Lemma* 7.1, *it is known that* 1 *is a simple eigenvalue of the monodromy operator* $U_{T,\tau}$ *and* $|\sigma_2(U_{T,\tau})| < 1$. *Then in some neighborhood of the point* $0 \in X_{0,\tau}$ *manifolds* $Y_{1,\tau}$ *and* $Y_{2,\tau}$ *are defined which are invariant with respect to the operator* K

and are tangent, respectively, to the subspaces $X_{1,\tau}$ *and* $X_{2,\tau}$. *Moreover*:

1) *If* $a_0 \in Y_{2,\tau}$, *then* $K^n a_0 \to 0$ *and* $\rho(\widetilde{N}_t x_0, C) \to 0$ *as* $t \to +\infty$; $x_0 = q_\tau + a_0$.

2) *There exists* $\varepsilon_0 > 0$ *such that for* $a_0 \notin Y_2$ *and some* $n_0 = n_0(a_0)$ *and* $t_0 = t_0(a_0)$

$$\|K^{n_0} a_0\| > \varepsilon_0, \qquad \rho(\widetilde{N}_{t_0} x_0, C) > \varepsilon_0.$$

3) *For any* $a_0 \in Y_1$ *the inverse mapping* $\widetilde{N}_s^{-1} = \widetilde{N}_{-s}$ $(s > 0)$ *and* $\rho(\widetilde{N}_t x_0, C) \to 0$ *as* $t \to -\infty$.

4) *If* $a_0 \notin Y_1$, *then there exists* $t_1 < 0$ *such that* $\rho(\widetilde{N}_{t_1}, x_0, C) > \varepsilon_0$.

Thus, under the conditions of Lemma 7.2 any trajectory with initial point in the set

$$Z_2 = \bigcup_{0 \le \tau \le T} (q_\tau + Y_{2,\tau}) = \bigcup_{0 \le \tau \le T} \widetilde{N}_\tau(q_0 + Y_{2,0})$$

tends to a cycle as $t \to +\infty$, while any trajectory with initial point in $Z_1 = \bigcup_{0 \le \tau \le T}(q_\tau + Y_{1,\tau})$ tends to a cycle as $t \to -\infty$. If the initial point x_0 is close to a cycle but does not lie either on Z_1 or on Z_2, then there exist $t_0 = t_0(x_0)$ and $t_1 = t_1(x_0) < 0$ such that the points of the trajectories $\widetilde{N}_{t_0} x_0$ and $\widetilde{N}_{t_1} x_0$ are located outside some fixed neighborhood of the cycle (these points are possibly not defiend at all). We thus obtain a rather complete description of the behavior of trajectories in a neighborhood of a cycle. We remark that Z_1 and Z_2 are invariant sets of the dynamical system $(\widetilde{N}_t, X)$.

In order to be able to apply Lemmas 7.1 and 7.2 to the Navier-Stokes equations it suffices to show that the vector-valued function

$$v(t) = \widetilde{N}_t(v(0) + a)$$

defined by (6.41) is strongly continuously differentiable with respect to t for any $t > 0$ and any a in some neighborhood of zero of the space H_1.

Suppose $\psi(t)$ is an infinitely differentiable function equal to 0 for $0 \le t \le \delta$ and to 1 for $t \ge 2\delta$; δ is an arbitrarily fixed positive number.

We introduce the vector-valued functions η_h and ζ_h by setting for any $h > 0$

$$\eta_h(t) = \psi(t)\zeta_h(t); \qquad \zeta_h(t) = \frac{1}{h}[v_h(t) - v(t)]; \qquad (7.40)$$
$$v_h = v(t+h).$$

it is clear that $\zeta_h \to dv/dt$ and $\eta_h \to dv/dt$ in the norm of the space $L_2((0,T), H)$ as $h \to 0$; this is obvious for smooth v, and in the general case it is easily proved by means of the Banach-Steinhaus theorem. From (6.1)–(6.3) we derive for η_h the following differential equation:

$$\frac{d\eta_h}{dt} + A_0\eta_h + K(v(t+h), \eta_h(t)) + K(\eta_h(t), v(t)) = \psi'\zeta_h. \tag{7.41}$$

Arguing as in Lemma 2.4, we show that, uniformly with respect to h and τ (τ is a fixed positive number, $2\delta < \tau$),

$$\|\eta_h\|_{H_2^\tau} \leq C, \tag{7.42}$$

where the constant C depends only on $\|v_0\|_{\widetilde{H}_2^T}$ and on τ. From (7.42) it follows that the family $\{\eta_h\}$ is weakly compact. We show that η_h converges weakly as $h \to 0$. Indeed, suppose $\eta_{hn} \to \eta$ weakly in H_2^τ for some sequence $h_n \to 0$. Passing to the limit in (7.41) with consideration of the imbeddings indicated in Lemmas 2.4 and 2.5, we obtain

$$\frac{d\eta}{dt} + A_0\eta + K^0(v, \eta) = \psi'\zeta. \tag{7.43}$$

Just as in the proof of Lemmas 2.3 and 2.4, we see that (7.43) together with the initial condition $\eta(0) = 0$ uniquely determines the vector-valued function $\eta \in H_2^\tau$.

We now recall that $\eta(t) = dv(t)/dt$ for $t \geq 2\delta$. Applying Lemma 2.1, we conclude that the vector-valued function v is differentiable with respect to t in H_1 for any $t > 0$. Continuity of the dependence of the derivative dv/dt for $t > 0$ on the initial vector a can be established as in Lemma 2.5. This completes the proof of Theorem 7.1.

§8. Damping of the leading derivatives

It is well known that for smooth data sufficiently regular generalized solutions of parabolic equations and the Navier-Stokes system improve their differential properties. If at the initial time the velocity field has discontinuities, then for $t > 0$ they vanish. Under the conditions of the stability theorems proved above, perturbations die out not only in the norm of L_p or $W_p^{(1)}$ but also in a stronger sense. For example, the following assertion holds.

THEOREM 8.1. *Suppose the boundary $S \in C^\infty$ and the periodic flow $v_0(t)$ is infinitely differentiable. Then under the conditions of Theorem* 3.1 *it is asymptotically η-stable (H_1, C^k) for any $k = 1, 2, \ldots$. Moreover, for the perturbation $u(t)$*

$$\|D_t^n u\|_{C^k(\Omega)} \to 0, \qquad t \to +\infty. \tag{8.1}$$

exponentially for any natural numbers n and k.

Proof. Because of the imbedding theorems it suffices to prove (8.1) with C^k replaced by $W_2^{(n)}(\Omega)$. We return to (7.43). In application to this equation the arguments carried out in the proof of Theorem 3.1 lead with consideration of (3.15) and (3.16) to the estimates

$$\|\eta(t)\|_{H_1} \leq Ce^{-\sigma_0 t}\|a\|_{H_1} \tag{8.2}$$

$$\int_0^\infty e^{2\sigma_0 t}\left(\left\|\frac{d\eta}{dt}\right\|_H^2 + \|A_0\eta\|_H^2\right) d\tau \leq C^2\|a\|_{H_1}^2, \tag{8.3}$$

which are satisfied for all small $a \in H_1$.

Since for $t \geq 2\delta$ the vector-valued function η coincides with dv/dt, from (8.2) we obtain the required estimate of the latter. Applying the coerciveness inequality for the steady-state problem, we estimate the norm of $D_x^2 u$ in $L_2(\Omega)$ and see that it tends to 0 exponentially as $t \to +\infty$, while for any fixed $t > 0$ and $a \to 0$ in H_1 it vanishes.

We further successively differentiate (7.43) with respect to t and estimate the leading derivatives in a similar manner; this concludes the proof of Theorem 8.1.

The conditions on the initial data in this theorem can be relaxed. For simplicity we restrict ourselves to steady flows.

Theorem 8.2. *Under the conditions of Theorem* 8.1 *a steady flow* v_0 *is exponentially* η*-stable* (S_p, C^k) *for* $p > 3$ *and any natural number* k, *and* (8.1) *holds.*

Proof. It suffices for the perturbation u to prove that for small $a \in S_p$

$$\|u(t)\|_{H_1} \leq C(t)\|a\|_{S_p}, \qquad t > 0. \tag{8.4}$$

Now from Theorem 2.3 of Chapter II it follows that $\|u(t)\|_{S_q}$ and $\|D_x u\|_{q(\Omega\times[\delta,\infty)}^L$ for $t > 0$, $\delta > 0$, and any $q > 1$ can be estimated in terms of $\|a\|_{S_p}$ if $p > 3$. Then $\|Ku\|_{L_q(\Omega\times[\delta,\infty))}$ can also be estimated in terms of $\|a\|_{S_p}$, and inequality (8.4) follows, for example, from Theorem 3.3 of Chapter I. Theorem 8.2 is proved.

Bibliography

1. S. Agmon, A. Douglis, and L. Nirenberg, *Estimates near the boundary for solutions of elliptic partial differential equations satisfying general boundary conditions*, I. Comm. Pure Appl. Math. **12** (1959), 623–727.

2. S. Agmon and L. Nirenberg, *Properties of solutions of ordinary differential equations in Banach space*, Comm. Pure Appl. Math. **16** (1963), 121–239.

3. I. P. Andreĭchikov and V. I. Yudovich, *Self-oscillating regimes branching from a Poiseuille flow in a two-dimensional channel*, Dokl. Akad. Nauk SSSR **202** (1972), 791–794; English transl. in Soviet Phys. Dokl. **17** (1972/73).

4. D. V. Anosov, *A multidimensional analog of a theorem of Hadamard*, Nauchn. Dokl. Vyssh. Shkoly Fiz.-Mat. Nauki **1959**, no. 1, 3–12. (Russian)

5. ———, *Geodesic flows on closed Riemannian manifolds of negative curvature*, Trudy Mat. Inst. Steklov. **90** (1967); English transl., Proc. Steklov Inst. Math. **90** (1967).

6. V. Arnold [V. I. Arnol'd], *Sur la géométrie différentielle des groupes de Lie de dimensions infinie et ses applications à l'hydrodynamique des fluides parfaits*, Ann. Inst. Fourier (Grenoble) **16** (1966), fasc. 1, 319–361.

7. ———, *Mathematical methods in classical mechanics*, "Nauka", Moscow, 1974; English transl., Springer-Veerlag, 1978.

8. V. G. Babskiĭ, *On "thresholds of instability" for the occurrence of convection*, Contemporary Questions of Hydrodynamics (Materials, Summer School, Kanev, 1965; V. S. Tkalich, editor), "Naukova Dumka", Kiev, 1967, pp. 325–330. (Russian)

9. M. Sh. Birman and M. Z. Solomyak, *Asymptotic properties of the spectrum of differential equations*, Itogi Nauki i Tekhniki: Mat. Anal., vol. 14, VINITI, Moscow, 1977, pp. 5–58; English transl. in J. Soviet Math. **12** (1979), no. 3.

10. A. P. Calderón and A. Zygmund, *On the existence of certain singular integrals*, Acta Math. **88** (1952), 85–139.

11. S. Chandrasekhar, *Hydrodynamic and hydromagnetic stability*, Clarendon Press, Oxford, 1981.

12. Yu. L. Daletskiĭ and M. G. Kreĭn, *Stability of solutions of differential equations in Banach space*, "Nauka", Moscow, 1970; English transl., Amer. Math. Soc., Providence, R. I., 1974.

13. A. Davey, *The growth of Taylor vortices in flow between rotating cylinders*, J. Fluid Mech. **14** (1962), 336–368.

14. V. A. Dikarev and V. I. Matsaev, *An exact interpolation theorem*, Dokl. Akad. Nauk SSSR **168** (1966), 986–988; English transl. in Soviet Math. Dokl. **7** (1966).

15. Nelson Dunford and Jacob T. Schwartz, *Linear operators*. Vol. I: *General theory*, Interscience, 1958.

16. ——, *Linear operators*. Vol. II: *Spectral theory*, Interscience, 1963.

17. F. R. Gantmakher, *The theory of matrices*, 2nd ed., "Nauka", Moscow, 1966; English transl. of 1st ed., Vols. 1, 2, Chelsea, New York, 1959.

18. F. R. Gantmakher and M. G. Kreĭn, *Oscillation matrices and kernels and small oscillations of mechanical systems*, 2nd ed., GITTL, Moscow, 1950; German transl., Akademie-Verlag, Berlin, 1960.

19. V. P. Glushko and S. G. Kreĭn, *Fractional powers of differential operators and imbedding theorems*, Dokl. Akad. Nauk SSSR **122** (1958), 963–966. (Russian)

20. I. Ts. Gokhberg [Israel Gohberg] and M. G. Kreĭn, *Introduction to the theory of linear nonselfadjoint operators in Hilbert space*, "Nauka", Moscow, 1965; English transl., Amer. Math. Soc., Providence, R.I., 1969.

21. Paul R. Halmos, *A Hilbert space problem book*, Van Nostrand, Princeton, N.J., 1967.

22. Einar Hille and Ralph S. Phillips, *Functional analysis and semi-groups*, rev. ed., Amer. Math. Soc., Providence, R.I., 1957.

23. M. W. Hirsch, C. C. Pugh, and M. Shub, *Invariant manifolds*, Lecture Notes in Math., vol. 583, Springer-Verlag, 1977.

24. Eberhard Hopf, *Statistical hydromechanics and functional calculus*, J. Rational Mech. Anal. **1** (1952), 87–123.

25. Lars Hörmander, *Estimates for translation invariant operators in L^p-spaces*, Acta Math. **104** (1960), 93–140.

26. V. P. Il'in, *On an imbedding theorem for a limit exponent*, Dokl. Akad. Nauk SSSR **96** (1954), 905–908. (Russian)

27. ——, *Some functional inequalities of the type of imbedding theorems*, Dokl. Akad. Nauk SSSR **123** (1958), 967–970. (Russian)

28. ——, *Some inequalities in function spaces and their application to the study of the convergence of variational processes*, Trudy Mat. Inst. Steklov **53** (1959), 64–127; English transl. in Amer. Math. Soc. Transl. (2) **81** (1969).

29. Daniel D. Joseph, *Stability of fluid motions*, Vols. I, II, Springer-Verlag, 1976.

30. M. V. Keldysh, *On the eigenvalues and eigenfunctions of certain classes of nonselfadjoint equations*, Dokl. Akad. Nauk SSSR **77** (1951), 11–14; English transl., Appendix to the English transl. of A. S. Markus, *Introduction to the spectral theory of polynomial operator pencils*, Amer. Math. Soc., Providence, R.I., 1988.

31. Al Kelley, *The stable, center-stable, center, center-unstable, unstable manifolds*, J. Differential Equations **3** (1967), 546–570.

32. Yu. S. Kolesov, *Study of stability of solutions of second-order parabolic equations in the critical case*, Izv. Akad. Nauk SSSR Ser. Mat. **33** (1969), 1356–1372; English transl. in Math. USSR Izv. **3** (1969).

33. A. I. Koshelev, *A priori estimates in L_p and generalized solutions of elliptic equations and systems*, Uspekhi Mat. Nauk **13** (1958), no. 4 (82), 29–88; English transl. in Amer. Math. Soc. Transl. (2) **20** (1962).

34. M. A. Krasnosel′skiĭ, *Topological methods in the theory of nonlinear integral equations*, GITTL, Moscow, 1956; English transl., Pergamon Press, Oxford, and Macmillan, New York, 1964.

35. ——, *Positive solutions of operator equations*, Fizmatgiz, Moscow, 1962; English transl., Noordhoff, 1964.

36. ——, *The operator of translation along the trajectories of differential equations*, "Nauka", Moscow, 1966; English transl., Amer. Math. Soc., Providence, R.I., 1968.

37. M. A. Krasnosel′skiĭ et al., Integral operators in spaces of summable functions, "Nauka", Moscow, 1966; English transl., Noordhoff, 1975.

38. M. G. Kreĭn, *Lectures on the theory of stability of solutions of differential equations in Banach space*, Izdat. Akad. Nauk Ukrain. SSR, Kiev, 1964. (Russian) (Revised version, [12]).

39. S. G. Kreĭn, *On functional properties of operators of vector analysis and hydrodynamics*, Dokl. Akad. Nauk SSSR **93** (1953), 969–972. (Russian)

40. ——, *Linear differential equations in Banach space*, "Nauka", Moscow, 1967; English transl., Amer. Math. Soc., Providence, R.I., 1971.

41. A. L Krylov, *Proof of the instability of a certain flow of viscous incompressible fluid*, Dokl. Akad. Nauk SSSR **153** (1963), 787–790; English transl. in Soviet Phys. Dokl. **8** (1963/64).

42. ——, *On the stability of a Poiseuille flow in a two-dimensional channel*, Dokl. Akad. Nauk SSSR **159** (1964), 978–981; English transl. in Soviet Math. Dokl. **5** (1964).

43. O. A. Ladyzhenskaya, *Mathematical questions in the dynamics of a viscous incompressible fluid*, 2nd rev. aug. ed., "Nauka", Moscow, 1970; English transl. of 1st ed., *The mathematical theory of viscous incompressible flow*, Gordon and Breach, New York, 1963 (rev. 1969).

44. O. A. Ladyzhenskaya, V. A. Solonnikov, and N. N. Ural′tseva, *Linear and quasilinear equations of parabolic type*, "Nauka", Moscow, 1967; English transl., Amer. Math. Soc., Providence, R.I., 1968.

45. C. C. Lin, *The theory of hydrodynamic stability*, Cambridge Univ. Press, 1955.

46. J. L. Lions, *Sur la régularité et l'unicité des solutions turbulentes des équations de Navier-Stokes*, Rend. Sem. Mat. Univ. Padova **30** (1960), 16–23.

47. L. A. Lyusternik and V. I. Sobolev, *Elements of functional analysis*, 2nd rev. ed., "Nauka", Moscow, 1965; English transl., Hindustan, Delhi, and Wiley, New York, 1974.

48. G. S. Markham and V. I. Yudovich, *Numerical investigation of the occurrence of convection in a fluid layer under the action of external forces periodic in time*, Izv. Akad. Nauk SSSR Mekh. Zhidk. Gaza **1972**, no. 3, 81–86; English transl. in Fluid Dynamics **7** (1972).

49. ——, *Onset of convection regimes with double period in a periodic field of external forces*, Zh. Prikl. Mekh. i Tekhn. Fiz. **1972**, no. 6, 65–70; English transl. in J. Appl. Mech. Tech. Phys. **13** (1972).

50. J. E. Marsden, H. McCracken, and G. Oster, *The Hopf bifurcation and its applications*, Springer-Verlag, 1976.

51. L. D. Meshalkin and Ya. G. Sinaĭ, *Investigation of the stability of a steady-state solution of a system of equations for the two-dimensional movement of a viscous incompressible fluid*, Prikl. Mat. Mekh. **25** (1961), 1140–1143; English transl. in J. Appl. Math. Mech. **25** (1961).

52. S. G. Mikhlin, *The problem of the minimum of a quadratic functional*, GITTL, Moscow, 1952; English transl., Holden-Day, San Francisco, Calif., 1965.

53. ——, *Multidimensional singular integrals and integral equations*, Fizmatgiz, Moscow, 1962; English transl., Pergamon Press, 1965.

54. Carlo Miranda, *Equazioni alle derivate parziali di tipo ellittico*, Springer-Verlag, 1955; English transl., 1970.

55. A. S. Monin and A. M. Yaglom, *Statistical hydromechanics*. Part I, "Nauka", Moscow, 1965; rev. aug. English transl., MIT Press, Cambridge, Mass., 1971.

56. Yu. I. Neĭmark, *On the existence and structural stability of invariant manifolds for point-to-point mappings*, Izv. Vyssh. Uchebn. Zaved. Radiofizika **10** (1967), 311–320; English transl. in Radiophysics and Quantum Electronics **10** (1967).

57. ——, *Integral manifolds of differential equations*, Izv. Vyssh. Uchebn. Zaved. Radiofizika **10** (1967), 321–334; English transl. in Radiophysics and Quantum Electronics **10** (1967).

58. V. V. Nemytskiĭ and V. V. Stepanov, *Qualitative theory of differential equations*, 2nd ed., GITTL, Moscow, 1949; English transl., Princeton Univ. Press, Princeton, N.J., 1960.

59. S. M. Nikol′skiĭ, *Approximation of functions of several variables and embedding theorems*, "Nauka", Moscow, 1969; English transl., Springer-Verlag, 1975.

60. Louis Nirenberg, *Remarks on strongly elliptic partial differential equations*, Comm. Pure Appl. Math. **8** (1955), 649–675.

61. Giovanni Prodi, *Qualche risultato riguardo alle equazioni di Navier-Stokes nel caso bidimensionale*, Rend. Sem. Mat. Univ. Padova **30** (1960), 1–15.

62. ——, *Teoremi di tipo locale per il sistema di Navier-Stokes e stabilità delle soluzioni stazionarie*, Rend. Sem. Mat. Univ. Padova **32** (1962), 374–397.

63. V. A. Romanov, *Stability of plane-parallel Couette flow*, Preprint No. 1, Inst. Problems Mech., Acad. Sci. USSR, Moscow, 1971. (Russian)*

64. D. H. Sattinger, *The mathematical problem of hydrodynamic stability*, J. Math. and Mech. **19** (1969/70), 797–817.

65. J. Schwartz, *A remark on inequalities of Calderon-Zygmund type for vector-valued functions*, Comm. Pure Appl. Math. **14** (1961), 785–799.

66. James Serrin, *A note on the existence of periodic solutions of the Navier-Stokes equations*, Arch. Rational Mech. Anal. **3** (1959), 120–122.

67. I. B. Simonenko, *Boundedness of singular integrals in Orlicz spaces*, Dokl. Akad. Nauk SSSR **130** (1960), 984–987; English transl. in Soviet Math. Dokl. **1** (1960).

Editor's note.* See also his papers of the same title in Dokl. Akad. Nauk SSSR **196 (1971), 1049–1051, and Funktsional. Anal. i Prilozhen. **7** (1973), no. 2, 62–73; English transls. in Soviet Phys. Dokl. **16** (1971/72) and Functional Anal. Appl. **7** (1973).

68. ——, *A study in the theory of singular integrals, boundary value problems for analytic functions, and singular integral equations*, Candidate's Dissertation, Razmadze Math. Inst. Acad. Sci. Georgian SSR, Tbilisi, 1961. (Russian)

69. ——, *Interpolation and extrapolation of linear operators in Orlicz spaces*, Mat. Sb. **63(105)** (1964), 536–553. (Russian)

70. L. N. Slobodetskiĭ, *Estimates in L_p of solutions of elliptic systems*, Dokl. Akad. Nauk SSSR **123** (1958), 616–619. (Russian)

71. V. I. Smirnov, *Course in higher mathematics*, Vol. IV, 3rd ed., GITTL, Moscow, 1958; English transl., Pergamon Press, Oxford, and Addison-Wesley, Reading, Mass., 1964.

72. P. E. Sobolevskiĭ, *Coercivity inequalities for abstract parabolic equations*, Dokl. Akad. Nauk SSSR **157** (1964), 52–55; English transl. in Soviet Math. Dokl. **5** (1964).

73. ——, *Study of Navier-Stokes equations by the methods of the theory of parabolic equations in Banach spaces*, Dokl. Akad. Nauk SSSR **156** (1964), 745–748; English transl. in Soviet Math. Dokl. **5** (1964).

74. M. Z. Solomyak, *Application of semigroup theory to the study of differential equations in Banach spaces*, Dokl. Akad. Nauk SSSR **122** (1958), 766–769. (Russian)

75. V. A. Solonnikov, *On estimates of Green's tensors for certain boundary value problems*, Dokl. Akad. Nauk SSSR **130** (1960), 988–991; English transl. in Soviet Math. Dokl. **1** (1960).

76. ——, *A priori estimates for certain boundary value problems*, Dokl. Akad. Nauk SSSR **138** (1961), 781–784; English transl. in Soviet Math. Dokl. **2** (1961).

77. ——, *Estimates for solutions of time-dependent Navier-Stokes equations*, Zap. Nauchn. Sem. Leningrad. Otdel. Mat. Inst. Steklov. (LOMI) 38 (1973), 153–231; English transl. in J. Soviet Math. **8** (1977), no. 4.

78. ——, *Estimates of the solutions of a time-dependent linearized system of Navier-Stokes equations*, Trudy Mat. Inst. Steklov. **70** (1964), 213–317; Enslish transl. in Amer. Math. Soc. transl. (2) **75** (1968).

79. ——, *On general boundary value problems for Douglis-Nirenberg elliptic systems*. I, II, Izv. Akad. Nauk SSSR Ser. Mat. **28** (1964), 665–706; Trudy Mat. Inst. Steklov. **92** (1966), 233–297; English transls. in Amer. Math. Soc. Transl. (2) **56** (1966) and Proc. Steklov Inst. Math. **92** (1966).

80. E. M. Stein and Guido Weiss, *An extension of a theorem of Marcinkiewicz and some of its applications*, J. Math. and Mech. **8** (1959), 263–284.

81. M. M. Vaĭnberg and V. A. Trenogin, *Theory of branching of solutions of nonlinear equations*, "Nauka", Moscow, 1969; English transl., Noordhoff, 1974.

82. I. I. Vorovich and V. I. Yudovich, *Steady flow of a viscous fluid*, Dokl. Akad. Nauk SSSR **124** (1959), 542–545; English transl. in Soviet Phys. Dokl. **4** (1959/60).

83. ——, *Steady flow of a viscous incompressible fluid*, Mat. Sb. **53(95)** (1961), 393–428. (Russian)

84. V. I. Yudovich, *Periodic motions of a viscous incompressible fluid*, Dokl. Akad. Nauk SSSR **130** (1960), 1214–1217; English transl. in Soviet Math. Dokl. **1** (1960).

85. ——, *Loss of smoothness of solutions of the Euler equations with time*, Dinamika Sploshnoĭ Sredy Vyp. 16 (1974), 71–78. (Russian)

86. ——, *Mathematical questions of the theory of stability of flows of a fluid*, Doctoral Dissertation, Inst. Problems Mech., Acad. Sci. USSR, Moscow, 1972. (Russian)

87. ——, *Some estimates connected with integral operators and with solutions of elliptic equations*, Dokl. Akad. Nauk SSSR **138** (1961), 805–808; English transl. in Soviet Math. Dokl. **2** (1961).

88. ——, *Some bounds for solutions of elliptic equations*, Mat. Sb. **59(101)** (1962), 229–244; English transl. in Amer. Math. Soc. Transl. (2) **56** (1966).

89. ——, *On an estimate for the solution of an elliptic equation*, Uspekhi Mat. Nauk **20** (1965), no. 2, (122), 213–219. (Russian)

90. ——, *Stability of steady flows of viscous incompressible fluids*, Dokl. Akad. Nauk SSSR **161** (1965), 1037–1040; English transl. in Soviet Phys. Dokl. **10** (1965/66).

91. ——, *Example of the generation of a secondary steady or periodic flow when there is loss of stability of the laminar flow of a viscous incompressible fluid*, Prikl. Mat. Mekh. **29** (1965), 453–467; English transl. in J. Appl. Math. Mech. **29** (1965).

92. ——, *Mathematical questions of hydrodynamical stability theory*, All-Union Inter-Univ. Conf. Application of Methods of Functional Analysis to the Solution of Nonlinear Problems, Abstracts of Reports, Baku, 1965, p. 25. (Russian)

93. ——, *Stability of convection flows*, Prikl. Mat. Mekh. **31** (1967), 272–281; English transl. in J. Appl. Math. Mech. **31** (1967).

94. ——, *Instability of parallel flows of a viscous incompressible fluid with respect to perturbations periodic in space*, Numerical Methods for Solving Problems of Mathematical Physics (Supplement to Zh. Vychisl.

Mat. i Mat. Fiz. **6** (1966), no. 4), "Nauka", Moscow, 1966, pp. 242–249. (Russian)

95. ——, *Secondary flows and fluid instability between rotating cylinders*, Prikl. Mat. Mekh. **30** (1966), 688–698; English transl. in J. Appl. Math. Mech. **30** (1966).

96. ——, *The bifurcation of a rotating flow of liquid*, Dokl. Akad. Nauk SSSR **169** (1966), 306–309; English transl. in Soviet Phys. Dokl. **11** (1966/67).

97. ——, *Generation of secondary steady and periodic regimes on loss of stability of a steady flow of a fluid*, Abstracts Brief Sci. Comm., Section 12, Internat. Congr. Math., Moscow, 1966, p. 57. (Russian)

98. ——, *Questions in the mathematical theory of stability of fluid flows*, Third All-Union Congr. Theoret. Appl. Mech., Abstracts of Reports, Akad. Nauk SSSR, Moscow, 1968, p. 330. (Russian)

99. ——, *An example of loss of stability and generation of a secondary flow in a closed vessel*, Mat. Sb. **74(116)** (1967), 565–579; English transl. in Math. USSR Sb. **3** (1967).

100. ——, *On the stability of forced oscillations of a fluid*, Dokl. Akad. Nauk SSSR **195** (1970), 292–295; English transl. in Soviet Math. Dokl. **11** (1970).

101. ——, *On the stability of self-oscillations of a fluid*, Dokl. Akad. Nauk SSSR **195** (1970), 574–576; English transl. in Soviet Math. Dokl. **11** (1970).

102. ——, *The onset of self-oscillations in a fluid*, Prikl. Mat. Mekh. **35** (1971), 638–655; English transl. in J. Appl. Math. Mech. **35** (1971).

103. ——, *Investigations of self-oscillations of a continuous medium that arise at loss of stability of a steady regime*, Prikl. Mat. Mekh. **36** (1972), 450–459; English transl. in J. Appl. Math. Mech. **36** (1972).

104. A. Zygmund, *Trigonometric series*, 2nd rev. ed., Vols. I, II, Cambridge Univ. Press, 1959.

105. M. A. Gol′dshtik and V. N. Shtern, *Hydrodynamic stability and turbulence*, "Nauka", Novosibirsk, 1977. (Russian)

106. Yu. S. Barkovskiĭ and V. I. Yudovich, *Spectral properties of a class of boundary value problems*, Mat. Sb. **114(156)** (1981), 438–450; English transl. in Math. USSR Sb. **42** (1982).

107. V. A. Solonnikov, *On the differentiability properties of the solution of the first boundary value problem for a time-dependent system of Navier-Stokes equations*, Trudy Mat. Inst. Steklov. **73** (1964), 221–291. (Russian)

108. M. A. Naĭmark, *Normed rings*, GITTL, Moscow, 1966; English transl., Noordhoff, 1959.

ABCDEFGHIJ – 89